Downstream

Adaptive Management of Glen Canyon Dam and the Colorado River Ecosystem

Committee on Grand Canyon Monitoring and Research
Water Science and Technology Board
Commission on Geosciences, Environment, and Resources
National Research Council

National Academy Press
Washington, D.C.

NATIONAL ACADEMY PRESS • 2101 Constitution Avenue, N.W. • Washington, DC 20418

NOTICE: The project that is the subject of this report was approved by the Governing Board of the National Research Council, whose members are drawn from the councils of the National Academy of Sciences, the National Academy of Engineering, and the Institute of Medicine. The members of the committee responsible for the report were chosen for their special competences and with regard for appropriate balance.

Support for this project was provided by the Bureau of Reclamation under Cooperative Agreement # 1425-98-FC-40-22700.

Cover: Photo of the Colorado River by David Rubin, U.S. Geological Survey, Menlo Park, CA.

Printed in the United States of America

COMMITTEE ON GRAND CANYON MONITORING AND RESEARCH

JAMES L. WESCOAT, JR. *(Chair)*, University of Colorado, Boulder
TRUDY A. CAMERON, University of California, Los Angeles
SUZANNE K. FISH, University of Arizona, Tucson
DAVID FORD, David Ford Consulting Engineers, Sacramento, California
STEVEN P. GLOSS, University of Wyoming, Laramie
TIMOTHY K. KRATZ, University of Wisconsin, Trout Lake Station, North Boulder Junction
WENDELL L. MINCKLEY, Arizona State University, Tempe
PETER R. WILCOCK, The Johns Hopkins University, Baltimore, Maryland

National Research Council Staff

JEFFREY W. JACOBS, Project Study Director
ANITA A. HALL, Project Assistant

The National Academy of Sciences is a private, nonprofit, self-perpetuating society of distinguished scholars engaged in scientific and engineering research, dedicated to the furtherance of science and technology and to their use for the general welfare. Upon the authority of the charter granted to it by the Congress in 1863, the Academy has a mandate that requires it to advise the federal government on scientific and technical matters. Dr. Bruce Alberts is president of the National Academy of Sciences.

The National Academy of Engineering was established in 1964, under the charter of the National Academy of Sciences, as a parallel organization of outstanding engineers. It is autonomous in its administration and in the selection of its members, sharing with the National Academy of Sciences the responsibility for advising the federal government. The National Academy of Engineering also sponsors engineering programs aimed at meeting national needs, encourages education and research, and recognizes the superior achievements of engineers. Dr. William A. Wulf is president of the National Academy of Engineering.

The Institute of Medicine was established in 1970 by the National Academy of Sciences to secure the services of eminent members of appropriate professions in the examination of policy matters pertaining to the health of the public. The Institute acts under the responsibility given to the National Academy of Sciences by its congressional charter to be an adviser to the federal government and, upon its own initiative, to identify issues of medical care, research, and education. Dr. Kenneth I. Shine is president of the Institute of Medicine.

The National Research Council was organized by the National Academy of Sciences in 1916 to associate the broad community of science and technology with the Academy's purposes of furthering knowledge and advising the federal government. Functioning in accordance with general policies determined by the Academy, the Council has become the principal operating agency of both the National Academy of Sciences and the National Academy of Engineering in providing services to the government, the public, and the scientific and engineering communities. The Council is administered jointly by both Academies and the Institute of Medicine. Dr. Bruce Alberts and Dr. William A. Wulf are chairman and vice chairman, respectively, of the National Research Council.

Preface

On March 26, 1996, the bypass tubes of Glen Canyon Dam were opened for the first experimental "controlled flood" in the Grand Canyon of the Colorado River, marking a dramatic physical start for an even broader Adaptive Management Program ("Program"). The Program aims to monitor and analyze the effects of dam operations on downstream resources in the Grand Canyon ecosystem and to use that knowledge to recommend to the U.S. Secretary of the Interior, on a continuing basis, adjustments intended to preserve and enhance downstream values.

Responsibility for scientific research and monitoring to support adaptive management rests with the Grand Canyon Monitoring and Research Center ("Center") in Flagstaff, Arizona. During the past two years, the Center has established headquarters and hired staff, worked with stakeholder groups (known as the Adaptive Management Work Group and the Technical Work Group), commissioned a "conceptual model" of the Grand Canyon ecosystem, established protocols for research funding, and let contracts for research and monitoring.

These actions have been guided in part by the Center's 1997 *Long-term Strategic Plan* (Center, 1997) which underwent initial revisions in 1998. As part of these revisions, the Center arranged for the National Research Council's (NRC) Water Science and Technology Board to review the Strategic Plan. Later in the year, the 1998 Draft Strategic Plan became a source of debate among stakeholder groups, and controversies are still being worked out.

The National Research Council appointed a special committee to assess the Strategic Plan from as many perspectives as seemed relevant to its roles in guiding this important experiment in United States environmental science and policy. This report documents our assessment of the Center's long-term strategic planning for monitoring and research in the Grand Canyon of the Colorado River, and is submitted with appreciation and constructive criticism. The Center's scientists have launched its research programs with energy, intelligence, and commitment. This committee's concerns range from the types of science and monitoring planned for the Grand Canyon, to the uses of scientific findings in the Adaptive Management Program, to the uses of advice from the Adaptive Management Program by the Secretary of the Interior and, ultimately, to the effects of that advice on Grand Canyon resources.

Challenges encountered in the Glen Canyon Environmental Studies (GCES), which preceded the current program, and in other adaptive management programs have special relevance for the Center's efforts. The Adaptive Management Program carries forward twelve years of work by the Glen Canyon Environmental Studies. This National Research Council (NRC) review continues over a decade (1985-1996) of prior NRC reviews of Glen Canyon Environmental Studies programs, the Glen Canyon Dam Environmental Impact Statement, and early plans for long-term monitoring.

Our report is titled *Downstream: Adaptive Management of the Glen Canyon Dam and the Colorado River Ecosystem* for three reasons. First, the Program's primary focus is literally on resources "downstream" of Glen Canyon Dam, a focus that remains contested in ways discussed in our report. In a figurative sense, adaptive management requires a "downstream" perspective beginning with hypothesized effects of dam-operation alternatives, followed by monitoring and research to test those hypotheses, and by further adjustments to dam operations. A downstream perspective requires a framework for envisioning ex ante courses of action that may be "adaptive" and for evaluating ex post the classes of outcomes that have or have not been adaptive. Third, "downstream" alludes to an earlier National Research Council report (1996b) on ecosystem management in the Columbia River basin entitled *Upstream: Salmon and Society in the Pacific Northwest*. Our report is briefer than *Upstream*, just as the Glen Canyon Dam Adaptive Management Program is more recent and more geographically focused than the Columbia River program. Nonetheless, the need for probing comparisons of adaptive management experiments underway in different regions of North America is one important conclusion of this report.

Our committee thanks the Center and its staff for their hospitality during site visits and for their open cooperation throughout the review process. The committee expresses special appreciation to David Garrett, the former chief who launched the Center and initiated the National Research Council review; Barry Gold, acting chief of the Center; Ruth Lambert, director of the Socioeconomic and Cultural Resources Programs; Mike Liszewski, director of the Information Technology Program; Ted Melis, director of the Physical Resources Program; Barbara Ralston, director of the Biological Resources Program; and all other Center scientists and staff. We also thank David Wegner, former chief of the Glen Canyon Environmental Studies, for speaking with the committee at its second meeting in Flagstaff, Arizona, in August 1998. As part of their review, committee members consulted with colleagues and stakeholders who offered useful insights and cautions that, collectively, helped guide our observations, evaluations, and recommendations. Anne Colgan, Ernest House, and Ann Huff of the University of Colorado gave helpful advice on the fields of strategic management and evaluation. William Clark, Harvard University; Kai Lee, Williams College; Steve Light, Institute for Agricultural and Trade Policy (formerly of the Minnesota Department of Natural Resources); Roger Pulwarty, NOAA; and John Volkman, Northwest Power Planning Council, generously shared their ideas about adaptive management.

As committee chair, I thank fellow committee members for their spirited contributions and thoughtful deliberation and exchange on interdisciplinary issues, as well as their written contributions to this report. Committee members tried out ideas, advancing some and dropping others as perspectives began to take shape on where the Center stands today and what a "long-term strategic plan" could and should entail.

Our committee owes special thanks to Jeffrey W. Jacobs of the National Research Council—first for directing the study on behalf of the Water Science and Technology Board, and second for his intellectual contributions to the review, especially on issues of water policy and adaptive environmental management. Anita Hall of the Water Science and Technology Board kept project communications, travel, and administration in order. Rhonda Bitterli provided thorough editorial advice on the committee's draft report.

This report has been reviewed by individuals chosen for their diverse perspectives and technical expertise, in accordance with procedures approved by the National Research Council's Report Review Committee. The purpose of this independent review is to provide candid and critical comments that will assist the institution in making the published report as sound as possible and to ensure that the report meets institutional standards for objectivity, evidence, and responsiveness to the study charge. The review comments and draft manuscripts remain confidential to protect the integrity of the deliberative process. We wish to thank the following individuals for their participation in the review of this report: Ellis Cowling, North Carolina State University (Emeritus); Tom Graff, Environmental Defense Fund, Oakland, California; Thomas Haan, Oklahoma State University; Duncan Patten, Arizona State University (Emeritus); Ronald Pulliam, University of Georgia; Jack Schmidt, Utah State University; Daniel Tarlock, Chicago-Kent College of Law; Henry Vaux, University of California, Oakland; John Warme, Colorado School of Mines; and Kenneth Weber, U.S. National Park Service. While the individuals listed above provided constructive comments and suggestions, it must be emphasized that responsibility for the final content of this report rests entirely with the authoring committee and the institution.

To all these persons we express our thanks. And to all the scientists and stakeholders concerned with the Grand Canyon—its waters, environment, and cultural significance—we direct this report with the hope that it helps advance the historic experiment in adaptive management that is underway.

James L. Wescoat Jr., Chair
University of Colorado at Boulder

Contents

APPENDIXES

Executive Summary

Glen Canyon Dam, authorized by the Colorado River Storage Project Act of 1956 and completed by the U.S. Bureau of Reclamation in 1963, spans the Colorado River just south of the Arizona–Utah border. Behind the dam, the waters of Lake Powell stretch upstream for 186 miles. Downstream, the Colorado River passes through a 15-mile stretch of Glen Canyon and the Glen Canyon National Recreation Area into Marble Canyon at Lee's Ferry, where it enters Grand Canyon National Park.

The river then flows 278 miles through Grand Canyon National Park before reaching Lake Mead, which is impounded behind Hoover Dam. Indian reservations, federal public lands, and private lands flank the Grand Canyon corridor. The Grand Canyon has deep cultural and ecological importance for numerous social groups and, as a World Heritage Site, it is important internationally and globally, as well. Flows through Glen Canyon Dam's eight hydroelectric turbines generate power for a multistate grid served by the Western Area Power Administration. Glen Canyon Dam and its operations have altered hydrologic and temperature regimes in ways that have dramatically transformed the Colorado River ecosystem.

Recognizing the "values for which Grand Canyon National Park and Glen Canyon National Recreation Area were established," the Grand Canyon Protection Act of 1992 (sec. 1802a) mandated an environmental impact statement and long-term monitoring of dam operation impacts on "resources of the Colorado River downstream of Glen Canyon Dam" (sec. 1801d). The final environmental impact statement was completed in March 1995. Nine alternatives, formulated through public input, technical data, interdisciplinary discussion, and professional judgment were selected for detailed study by an interagency environmental impact statement team. The preferred alternative—"modified low fluctuating flows"— specified minimum and maximum flow rates and ramping rates and provided for controlled floods to protect, enhance, and restore downstream resources.

The Glen Canyon Dam Environmental Impact Statement identified a set of expected benefits associated with the preferred alternative, but it also recognized scientific uncertainties regarding the extent and ways in which those benefits could be achieved. The preferred alternative was and is an experiment. To implement the experiment, and adjust it based on long-term monitoring and research, the Glen Canyon Dam Environmental Impact Statement recommended a program of "adaptive management." Though the concept is still evolving, adaptive management employs scientific monitoring and research to measure and explain the effects of management actions. Results of monitoring and research are then used to adjust future management strategies. In addition to the mandates of the Grand Canyon Protection Act, decisions regarding Glen Canyon Dam operations are constrained by an array of legal requirements, including the "Law of the River," the Endangered Species Act, and federal trust responsibilities to Indian tribes.

On October 8, 1996, the U.S. Secretary of the Interior signed the Record of Decision that established the Adaptive Management Program ("Program"), which is composed of the following:

(1) the Secretary of the Interior's designee,
(2) the Adaptive Management Work Group (AMWG),
(3) the Technical Work Group (TWG),
(4) independent review panels, and
(5) the Grand Canyon Monitoring and Research Center (GCMRC or "Center").

The Grand Canyon Monitoring and Research Center began long-term planning at its inception and, in May 1997, produced a *Long-Term Monitoring and Research Strategic Plan* that was adopted by stakeholder groups (the Adaptive Management Work Group and the Technical Work Group) later that year. The Center then requested the National Research Council's (NRC) Water Science and Technology Board to evaluate this plan. The National Research Council committee was asked to address two main questions and five related questions:

1. Will the Long-Term Strategic Plan be effective in meeting requirements specified in the Grand Canyon Protection Act, the final Glen Canyon Dam Environmental Impact Statement, and Record of Decision?

a. Does the Long-Term Plan respond to the new adaptive management process called for by the Grand Canyon Protection Act and Glen Canyon Dam Environmental Impact Statement? Is the Grand Canyon Monitoring and Research Center functioning effectively in the Adaptive Management Program, especially regarding incorporation of all stakeholder objectives and information needs in the planning process?

b. Does the Long-Term Plan incorporate past research knowledge in developing new monitoring and research directions?

c. Has the Center appropriately addressed past reviews of Glen Canyon Environmental Studies programs in formulating new research directions?

2. Characterize weaknesses of the Long-Term Plan and recommend short and long-term science elements to the Grand Canyon Monitoring and Research Center to address identified weaknesses.

a. What weaknesses exist in the Long-Term Plan, and how do these weaknesses affect the potential effectiveness of the overall science program?

b. What science elements are necessary to correct specific plan weaknesses?

During the latter part of this committee's review, the Center's Strategic Plan was revised and then split into three documents, which are yet to be completed. Mindful of the plans' evolving nature, this report encompasses the 1997 Strategic Plan (still in effect); the 1998 Strategic Plan (a revision of the 1997 Plan); and subsequent developments through April 1999. In some cases, the committee identified specific science elements for improving Center programs. In other cases, guidance is offered at a general level. In yet other instances, solutions were not immediately clear and will have to be addressed by the Center and

Adaptive Management Program stakeholder groups over the long term and with use of the Strategic Plan. Our recommendations are organized under three broad headings: Strategic Planning and Adaptive Management Issues; Science Program Issues; and Organizational and Budget Issues.

STRATEGIC PLANNING AND ADAPTIVE MANAGEMENT ISSUES

Strategic Planning

While the Center clearly recognizes the important links between strategic planning and adaptive management, four strategic aspects of the Plan need clarification: priorities for the next five years; geographic scope; decadal time scales; and the public significance of science-based adaptive management.

Strategic Priorities

The Strategic Plan does not identify the key strategic challenges that must be addressed in the next five years. For example, the main challenge in 1996 was to establish the Center and the Adaptive Management Program. The 1998 Plan has elements of a "problem statement" in its section on science needs and chapter on the philosophy of monitoring, but that chapter is more a list of factors to consider than strategic challenges to address.

The Strategic Plan should identify strategic priorities for the next five years, building explicitly upon experience gained during the past two years.

Geographic Scope of Center Programs

The 1998 Strategic Plan described the Program's geographic scope as extending upstream into the forebay of Lake Powell, downstream to the western boundary of Grand Canyon National Park, and laterally to the elevation of maximum regulated discharge and the inundated area for annual predam peak flows of 90,000 cubic feet per second (cfs). It wisely left open possibilities for selected studies in lateral areas associated with higher flows, Lake Powell, tributary watersheds, comparable river reaches elsewhere in the basin, and Lake Mead. That openness and its potential budgetary implications became a source of stakeholder debates.

The Center nonetheless successfully negotiated a five-year monitoring plan for Lake Powell water quality parameters relevant to dam operations; awarded a research contract on archaeological site erosion with control sites upstream in Cataract Canyon; and collaborated with the National Oceanic and Atmospheric Administration on a study of El Niño's implications for dam operations and downstream resources. These activities point toward sound ways to *manage* geographic scope that should be incorporated in the Strategic Plan.

Rigid definitions of geographic scope will not serve the Adaptive Management Program well. After clearly defining the Program's geographic focus, decisions about geographic linkages with adjacent areas and larger scales should be made on a case-by-case basis, considering ecosystem processes, management alternatives, funding sources, and stakeholder interests.

Decadal Time Scales

When discussing time scales, the Strategic Plan does not mention decadal and multidecadal periods relevant for ecosystem monitoring and research. The multidecadal life spans and population dynamics of fish species (e.g., humpback chub, razorback and flannelmouth suckers) bear greatly upon monitoring program design. Social values and institutions also change over time scales of decades. A long-term strategic plan must, by definition, consider medium- and long-term ecological and social processes.

The Strategic Plan should explicitly indicate how the five-year planning time frame relates to multidecadal ecological and social processes that are the real subjects of monitoring and research.

Public Significance of Science-Based Adaptive Management

The Center is responsible for addressing growing public policy interests in science-based approaches to adaptive management—

interests embodied in the Grand Canyon Protection Act, the Glen Canyon Dam Environmental Impact Statement, and the Record of Decision.   The Adaptive Management Program is a science-policy experiment of local, regional, national, and international importance.

The Strategic Plan should explicitly recognize and speak to public interests in Grand Canyon monitoring and research and should anticipate programs of public education, outreach, and involvement.

Adaptive Management

Although Center scientists have good working knowledge of theories and practices of adaptive management, six key aspects of its application to Glen Canyon Dam and the Grand Canyon ecosystem remain unclear.   These include: the definition of and roles in the Adaptive Management Program; the core adaptive management experiment; issues of "vision"; management objectives and information needs; a scientific basis for trade-off analysis and decision support systems; and independent scientific review.

Definition of and Roles in the Adaptive Management Program. The 1997 Strategic Plan defined adaptive management as follows: "Adaptive management begins with a set of management objectives and involves a feedback loop between the management action and the effect on that action on the system.... It is an iterative process, based on a scientific paradigm that treats management actions as experiments subject to modification, rather than as fixed and final rulings, and uses them to develop an enhanced scientific understanding about whether or not and how the ecosystem responds to scientific management actions" (Center, 1997). It is not clear whether this definition is widely shared or whether stakeholders and scientists have similar interpretations, particularly as it applies to Glen Canyon Dam operations and Grand Canyon ecosystem management.   As the use of ecosystem science develops in the Adaptive Management Program and as a vision for downstream resources becomes clearer, adaptive management may evolve into a program of ecosystem management.

The operational roles of scientific monitoring and research, and of the Center itself, remain unclear. A balance has not yet been reached among the Center's roles in conducting science programs, managing contracts, managing information systems, responding to stakeholder requests, and synthesizing and communicating monitoring and research results. At this writing, the Technical Work Group was drafting a Guidance Document to clarify those roles. The Center should be involved in the process of clarifying responsibilities to fully represent the functions of scientific monitoring and research.

The Center and the Adaptive Management Program stakeholders should work toward a common definition of adaptive management for the Grand Canyon ecosystem. The Center's various responsibilities in the Adaptive Management Program should be reviewed and clarified.

The Core Adaptive Management Experiment. The Strategic Plan describes the general role of experimentation in adaptive management but does not specifically define the core experiment with dam operations, as that experiment is specified in the Record of Decision. Clear articulation of this core experiment is needed to guide science and monitoring and to focus discussions among stakeholders. The Strategic Plan also could and should treat stakeholders' uses of monitoring and research results as scientific experiments.

The Center should clearly articulate the core adaptive management experiment in the Grand Canyon and, in particular, the hypothesized relations between dam operations, ecosystem responses, cultural effects, and trade-offs among consequent socioeconomic effects.

Issues of "Vision." Neither the Strategic Plan nor stakeholder groups have articulated a vision for the future state of the Grand Canyon ecosystem. A table of expected benefits from the preferred alternative in the Glen Canyon Dam Environmental Impact Statement represents a first step, but it is acknowledged to represent a compromise that is not internally consistent, optimal, or readily visualized (U.S. Bureau of Reclamation, 1995; see Appendix D of this report).

As the Adaptive Management Program is in its formative stages, it may be unrealistic to expect stakeholders and scientists to have agreed upon a common vision. The current pluralistic situation, however, constrains the Center's ability to synthesize scientific information and to employ certain scientific methods (e.g., rule-based simulation, optimization). Over time, as trade-offs are addressed among competing objectives and as a broader range of alternatives is examined, efforts to formulate a common vision for the Grand Canyon ecosystem may prove useful.

The Strategic Plan should recognize limitations of the current, pluralistic management situation. It should present a strategy for moving toward a set of common objectives and reference conditions for monitoring and research over the next five years.

Management Objectives and Information Needs. According to the 1998 Strategic Plan, stakeholder-defined management objectives (MOs) are intended to "define measurable standards of desired future conditions which will serve as objectives to be achieved by all stakeholders in the Glen Canyon Dam Adaptive process." Information needs (INs) "define the specific scientific understanding required to obtain specified management objectives."

The 1998 Strategic Plan listed 36 management objectives and 176 information needs. Some are hard to understand, redundant, or not measurable; and some information needed for ecosystem and socioeconomic analysis is not included. There are few cases of cross-program linkages. The lack of a clear and coherent set of management objectives and information needs makes it difficult to design or test adaptive management experiments.

The Center or a newly designated senior scientist, or both, should work with the Technical Work Group to develop a revised set of management objectives and information needs. These should be linked with testable hypotheses and situated within an internally consistent understanding of the ecosystem, for consideration by the Adaptive Management Work Group.

Scientific Basis for Trade-off Analysis and Decision Support Systems. Adaptive management ultimately involves trade-offs among competing objectives. The Strategic Plan concentrates on quantifying physical, biological, cultural, and conventional financial consequences of dam operations. It sidesteps the final, equally essential step of articulating scientific criteria for guiding choices among competing objectives that "protect, mitigate adverse impacts to, and improve the values" identified in the Grand Canyon Protection Act. While those choices rest ultimately with the Secretary of the Interior, the Center should work with stakeholder groups to develop decision support systems that aid those efforts.

It should be recognized that adaptive management for the Grand Canyon ecosystem will require trade-offs among management objectives favored by different stakeholder groups. It is recommended that the Adaptive Management Work Group begin to consider mechanisms for equitable weighting of competing interests and that the Center begin to develop decision support systems and methods. The Center's revised Strategic Plan should include a strategy for scientific evaluation of management alternatives, both in terms of ecological outcomes and satisfaction of stakeholder groups. The Strategic Plan should include a strategy for using—and evaluating the usefulness of—new scientific information in testing management alternatives, including their impacts on the welfare of different stakeholder groups.

Independent Science Review. Three levels of independent review are appropriate for the Center: (1) external review of research proposals and reports, (2) review of individual resource programs, and (3) broad programmatic review of the Center and Adaptive Management Program. This final level needs further attention.

The Strategic Plan calls for a Science Advisory Board that could be called upon for programmatic review, but two published requests for nominations have been unsuccessful. A discussion paper dated March 17, 1998, recommended that a Science Advisory Board be constituted as an official subcommittee of the Adaptive Management Work Group and that it be instructed to "not review, interpret, or otherwise evaluate public policy decisions associated with the Glen Canyon Dam Adaptive Management Program and activities of the AMWG, the TWG, or individual member agencies." These constraints would limit credible, independent review.

To ensure credible and independent programmatic review, the tasks and constraints of the Science Advisory Board should be redefined. It should not be a subcommittee of the Adaptive Management Work Group. Formal constraints should not be placed on issues that the Science Advisory Board would deem relevant to its charge.

SCIENCE PROGRAM ISSUES

The Strategic Plan describes the Center's commitment to ecosystem science and monitoring, and its five resource programs—physical resources, biological resources, cultural resources, socio-economic resources, and information technology.

Ecosystem Science and Monitoring

The main ecosystem science component of the strategic plans has been the development of a conceptual model. The model, along with a 1999 Colorado River Ecosystem Science Symposium, is helping integrate the scientific thinking of Center staff and other scientists working in the Grand Canyon. Although central to the Center's mission, a well-defined monitoring program has not yet been articulated.

Development and implementation of a detailed, long-term monitoring program should be a high priority for the Center. The monitoring program should be framed within a long-term perspective (in increments of five, ten, and more years).

Physical Resources Program

The Center's Physical Resources Program is well integrated and is actively engaged in the Adaptive Management Program. Much of the work is organized within a sediment budget model, which serves to identify parts of the system where additional study is needed. Future physical studies should:

- **Complete a sediment budget with acceptable levels of accuracy.**
- **Develop a long-term sand budget for Glen and Marble canyons and track the transport of tributary sediment inputs through Marble Canyon.**
- **Evaluate potential sediment conservation effects of beach/ habitat-building flows for larger flows and in all months of the year.**

Biological Resources Program

The Strategic Plan needs to provide a thorough synopsis of previous biological research in the Grand Canyon. The biological resources section of the Plan discusses broad resource management and monitoring principles but provides little specific indication of how they relate to the Grand Canyon ecosystem. One limitation, noted by a previous National Research Council committee (1996a), is the lack of linkages and lack of consistency between the Biological and Physical Resources Programs.

Another limitation is an emphasis on a few species rather than on communities or ecosystems. This is evidenced in management objectives and information needs focused on habitat enhancement and maintenance for listed or candidate species, and on compliance with recovery stipulations to prevent future listing or jeopardy opinions. Future biological research should:

- **Include a detailed review of existing knowledge about biological species and ecosystems in order to promote scientific reconstruction of biological changes in the Grand Canyon.**
- **Move away from a species-oriented emphasis toward broader monitoring and research on communities and ecosystems.**
- **Address biological aspects of temperature-control experiments involving the proposed selective withdrawal structure at Glen Canyon Dam.**

Sociocultural Resources Program

The 1998 Strategic Plan combined cultural and socioeconomic resources under a single heading. Such integration is promising, as it could facilitate comparisons of the effects of dam operations on different social groups. However, the Center's limited commitment to socioeconomic analysis, the magnitude of its responsibilities under the Cultural Resources Program, and limited staffing levels of these programs are troubling.

Cultural Resources

The Cultural Resources Program is the Center's third largest program (after the Biological Resources and Physical Resources Programs) and its most complex. It includes monitoring and research activities, cooperative and individual tribal projects, and coordination with a Programmatic Agreement between the U.S. Bureau of Reclamation, the U.S. National Park Service, the Advisory Council on Historic Preservation, and nine tribes (although two tribal groups have not signed).

The Center's Cultural Resources Program displays clearly defined relationships between management objectives, information needs, and proposed activities. Archaeological and anthropological elements of the Strategic Plan are integrated with the Physical Resources Program, although less so with ecosystem studies or conceptual modeling. While progress has been made in coordinating Center and Programmatic Agreement activities, the broader challenges of coordinating them with tribal projects remain. The apparent lack of resources for full tribal participation is another concern.

- **Coordinating cultural and socioeconomic programs is a worthwhile venture that should be tested and given sufficient resources. Further coordination of existing Cultural Resources subprograms is also needed.**
- **The Cultural Resources Program should look forward to including a wider range of social groups and to recognizing that archaeological evidence and ethnographic perspectives offer valuable insights on adaptive environmental management in the Grand Canyon.**
- **Resources must be secured for full tribal participation in all aspects of monitoring, research and communication in the Adaptive Management Program, without reducing other components of the Cultural Resources Program.**

Socioeconomic Resources

The 1998 Strategic Plan limits consideration of "economics" to recreation and hydropower. Limiting the scope of "economics" to two narrowly defined sources of benefits and costs associated with management decisions is disproportionate with the level of scrutiny of physical and biological effects associated with alternative management strategies. Aside from one useful project on recreation, no socioeconomic research on the effects of river management or other uses of the Grand Canyon is planned. This strategy fails to anticipate the types of social scientific knowledge needed for adaptive management.

A chain of analysis is necessary to inform good policy decisions. Managers first need to know how a change in flow regime will affect physical characteristics in the Grand Canyon; the effects of physical changes on flora and fauna then need to be quantified; and managers must then evaluate the impacts of these changes on the welfare of all stakeholder constituencies. The Center's budget and activities are devoted mainly to these first two points. The third is represented only by an incomplete measure of recreational user values and by the market costs of hydropower.

Knowledge of resource values to different constituencies and of how these change over time is important for effective resource management. Center staff should be familiar with current techniques for establishing social values for ecosystem services and should acquire expertise in these topics. One person currently manages the entire Sociocultural Resources Program. It is unrealistic to expect one person to effectively implement and coordinate the complex and diverse topics of cultural resources, tribal programs, and socioeconomics.

- **The Center should begin to develop internal expertise in techniques for nonmarket valuation of ecosystems and their services.**
- **The Strategic Plan should seek to understand not simply the range of preferences and activities of users of Grand Canyon resources, but also the degree to which ecosystem features and activities are valued.**
- **Sources of funding for original research devoted to measuring Grand Canyon ecosystem values should be sought, using a fully representative scientific sample of all stakeholders.**
- **Research is needed to develop a socioeconomic and cultural basis for evaluating the outcomes of adaptive management experiments based on meaningful comparison of the Grand Canyon's diverse resources.**

Information Technology Program

The Information Technology Program is functioning effectively in a support capacity; it is not a research or monitoring program. This role is appropriate, as the program supports the science and does not drive it. The program's goal is "to satisfy the information needs of stakeholders, scientists, and the public relative to the Colorado River ecosystem" (Center, 1998b). To fulfill this goal, the program has three tasks: (1) archiving and delivering scientific data and other information, (2) providing technology-based solutions to data collection, manipulation, and analysis, and (3) providing support in areas of computers, surveying, and geographic information systems. With some modifications, this program could better serve the needs of the stakeholders, scientists, and the public.

- **Information users should be surveyed to determine their information needs.**
- **Data archiving should be assigned a higher priority.**
- **Data and information delivery should be expanded and accelerated through the World Wide Web.**
- **Computer system administration should be managed independently of other Information Technology Program activities.**
- **The Center should begin to plan and develop a computerized decision support system(s).**

ORGANIZATIONAL AND BUDGET ISSUES

When assessing how the Center is functioning in the Adaptive Management Program, the committee encountered four main issues that are not fully addressed in the Strategic Plan: the roles of the Center; its institutional home; its structure and staffing; and its budget and funding.

Roles of the Center

The Center has been expected to plan research and monitoring activities and to facilitate many Technical Work Group and Adaptive Management Work Group activities. This is contrary to a model wherein these two work groups create a vision for the state of the ecosystem and attendant management objectives and informational needs, and the Center addresses them with a monitoring and research program.

The Center has been responsive to stakeholder requests, expending considerable effort at the likely expense of monitoring and research programs. In the process, however, the Center may become a subservient junior partner in the Program. The Technical Work Group seems to have emerged as the Adaptive Management Work Group's implementation arm and exerts decision-making powers over the Center's plans and budgets. These evolving relationships may constrain the Center's ability to fulfill its monitoring and research requirements.

The operational relationships and responsibilities of the Adaptive Management Program should be reviewed and reconsidered. Disproportionate oversight is presently exerted over governance and conduct of Center activities.

The Center's Institutional Home

The Center was temporarily formed under the Office of the Assistant Secretary of the Interior for Water and Science. This arrangement was helpful in facilitating research and monitoring activities and establishing a degree of independence for the Center. There remains, however, a high degree of interdependence between the Center and various Adaptive Management Program participants. For example, the Center uses U.S. Geological Survey (USGS) facilities, and it uses payroll and contractual services of the U.S. Bureau of Reclamation.

Based on three screening criteria, several alternatives for the Center's administative home have been considered. These include the U.S. Bureau of Reclamation, the U.S. Geological Survey, and the U.S. National Park Service, as well as maintaining the current interagency arrangement. Other alternatives that may have been considered include a university, an independent science organization such as the Smithsonian Institution, or a new interagency arrangement. All of these possibilities contain strengths and weaknesses. This review and previous National Research Council reports on institutional and administrative issues in the Glen Canyon Environmental Studies indicate that the following criteria, which resemble but extend beyond the screening criteria mentioned above, may be important in decisions regarding the Center's institutional home:

- **The Center should be housed within a premier science organization that has a commitment to physical, biological, and social science inquiry.**
- **The institutional home should enable the Center to work effectively with all Grand Canyon and Glen Canyon Dam management agencies.**
- **The institutional home should enable the Center to communicate scientific program issues and results directly with a management team at the Assistant Secretary level in the Department of the Interior.**
- **The Center should be independent from any single stakeholder management organization within the Adaptive Management Work Group.**

None of the arrangements currently considered perfectly satisfies all these criteria. The committee recommends that an institutional design, addressing institutional constraints and weaknesses related to these criteria, be part of a proposal for locating the Center within the U.S. Department of the Interior.

Center Structure and Staffing

With the retirement in 1998 of its first chief, the Center lost its most senior person. Previous National Research Council reviews called for and contributed to the appointment of a part-time senior scientist within the Glen Canyon Environmental Studies. The senior scientist was responsible for program design and synthesis and ensuring that these efforts fit both the ecosystem science paradigm and stakeholder needs.

The Center would benefit from the addition of a senior scientist, who would work with the stakeholder groups and Center staff to help clarify information needs and envision adaptive management experiments. The Center should also add an adaptive management specialist. This person would help articulate the links between scientific research and adaptive management experiments and their relations to policy recommendations for Grand Canyon ecosystem management. There also appear to be significant staffing needs in the Physical Resources, Cultural Resources, and Socioeconomic Resources programs.

A senior scientist and an adaptive management specialist should be appointed to the staff of the Center. Additional staff and

associated budget allocations seem warranted for the existing Physical Resources, Cultural Resources, and Socioeconomic Resources programs.

Center Budget and Funding

The Center's budget for monitoring and research is currently funded through proceeds from hydropower sales of the Western Area Power Administration. Although reasonable for core monitoring and research, there may be long-term disadvantages in drawing upon a single source of funding for all Center programs. It is thus recommended that the U.S. Department of the Interior:

Consider using hydropower revenues to support core research, monitoring, and Adaptive Management Program activities mandated by the Glen Canyon Dam Environmental Impact Statement, the Record of Decision, and the Grand Canyon Protection Act (at full funding levels envisioned for the next five years and beyond). Supplemental budgets for additional activities could be developed from U.S. Department of the Interior agencies, other federal agencies, and foundation sources.

In summary, this committee was impressed by the Center's strategic planning efforts to date. It is hoped that the recommendations in this report contribute to revised strategic plans, for the Center and the Adaptive Management Program, that fulfill the aims and requirements of the Grand Canyon Protection Act, the Glen Canyon Dam Environmental Impact Statement, and the Record of Decision.

1

Introduction

ORIGINS OF THE LONG-TERM STRATEGIC PLAN

A challenge in writing about the Glen Canyon Dam and Grand Canyon riverine ecosystem (Figure 1.1) is deciding where and how to begin. This review of the Grand Canyon Monitoring and Research Center's Strategic Plan (see http://www.gcmrc.gov) focuses on documents prepared between May 1997 and March 1999. The roots of these plans and associated programs, however, extend much deeper. They stem from the dramatic effects of construction, closure (in 1963), and subsequent operations of Glen Canyon Dam in one of the world's more beautiful landscapes. They have been shaped by lessons drawn from the Glen Canyon Environmental Studies program, which began to define the dam's impacts on the Grand Canyon ecosystem. They reflect the evolving "Law of the River" (the collection of compacts, statutes, judicial decisions, and regulations regarding Colorado River basin water), the changing roles of modern science in the Grand Canyon since John Wesley Powell's expeditions, and centuries of Native American experience in and knowledge of the Grand Canyon. While mindful of these roots, our review begins with formative events in the record of the Glen Canyon Environmental Studies, a program that extended from 1982 to 1996. It is not possible to clearly understand the current Strategic Plan or the debates surrounding it without this historical perspective.

The National Research Council (NRC) reviewed the first phase of the Glen Canyon Environmental Studies in 1986, concluding that "It

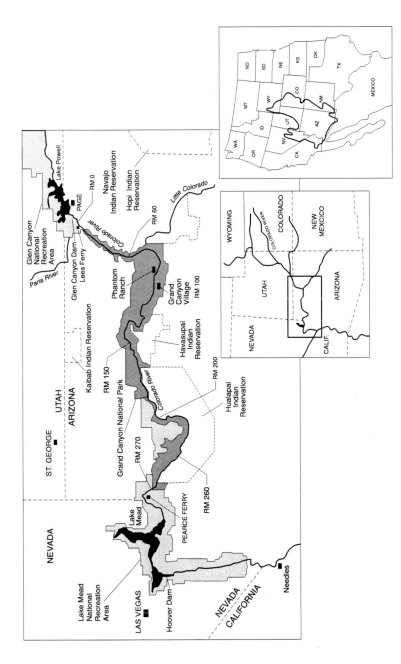

FIGURE1.1 Grand Canyon River Ecosystem and Colorado River Basin (Inset).
SOURCE: U.S. Bureau of Reclamation (1995).

cannot be stressed too strongly that detailed understanding of the Grand Canyon Ecosystem requires a well-planned monitoring program" (NRC, 1987, p. 78). Although long-term monitoring was envisioned by the Glen Canyon Environmental Studies, no monitoring plans were adopted in GCES Phase I (1982–1987). Further, the early stages of Phase I were disrupted in 1983 by uncontrolled flooding in the Grand Canyon. Research in GCES Phase II (1987–1996) was originally based on an ecosystem approach structured around specific hypotheses about the environmental and social effects of Glen Canyon Dam operations. This work, however, was disrupted by the immediate needs for data required to prepare the Glen Canyon Dam Environmental Impact Statement. Research flows were authorized to accelerate data acquisition and, following this, interim flows were applied to protect downstream resources until the Glen Canyon Dam Environmental Impact Statement was completed in 1995.

At the request of the Glen Canyon Environmental Studies' senior scientist, and in cooperation with the National Research Council, a workshop on long-term ecosystem monitoring was convened in 1992 in Irvine, California (NRC, 1992). A plan—Long-Term Monitoring in Glen and Grand Canyon: Response to Operations of Glen Canyon Dam (Patten, 1993)—was drafted, as required by the Grand Canyon Protection Act (Appendix A). This was the first step toward a future strategic plan. A National Research Council committee criticized the draft monitoring plan for neglecting the role of research, failing to estimate the likely costs of monitoring, not specifying the frequency and methods of monitoring, omitting information on administration and management, and not being clearly written (NRC, 1994). These criticisms, along with pressures from the U.S. Congress and the U.S. Department of the Interior for timely completion of the Glen Canyon Dam Environmental Impact Statement, reduced the momentum of long-term planning efforts.

The Final Environmental Impact Statement (U.S. Bureau of Reclamation, 1995) examined nine dam-operation alternatives, including the preferred "modified low fluctuating flow" (MLFF) alternative. Long-term monitoring and research was a common element for all alternatives, and was situated within a broader Adaptive Management Program ("Program") consisting of five organizational participants (Figure 1.2):

1. The Secretary's Designee – a person designated by the

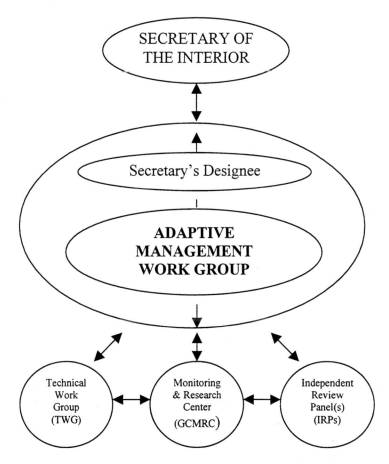

FIGURE 1.2 Organizations in the Adaptive Management Program.
SOURCE: Center (1997).

Secretary of the Interior to facilitate the Adaptive Management Program.

2. Adaptive Management Work Group (AMWG) – a federal advisory committee representing various stakeholder groups and meeting biannually on issues of policy (Appendix B).

3. Technical Work Group (TWG) – a federal advisory committee appointed by Adaptive Management Work Group members to address technical aspects of resource management (Appendix C).

4. Grand Canyon Monitoring and Research Center (GCMRC, or "Center") – a science center created in November 1995 to administer

monitoring and research needed by the Adaptive Management Program.

5. Independent Review Panels – panels established to provide independent review of the Center's scientific programs and documents.

The Secretary of the Interior's Record of Decision (ROD) on October 8, 1996 established the Adaptive Management Program and a modified version of the preferred alternative. Operating limits associated with the Record of Decision are listed in Table 1.1, and the entire Record of Decision is included as Appendix D.

To determine when to release a "beach/habitat-building flow" (described in the Record of Decision and based, in part, on the Colorado River Basin Project Act of 1968, sec. 602) the Adaptive Management Work Group adopted the following "hydrologic triggering criteria" (Adaptive Management Work Group, minutes of January 15, 1998 meeting):

1. If the January 1 forecast for the January-July unregulated spring runoff into Lake Powell exceeds 13 million acre-feet (about 140 percent of normal), assuming that Lake Powell is at approximately 3678 feet elevation (21.5 million acre-feet capacity), or

2. Any time a January-July Lake Powell inflow forecast would require a power-plant monthly release greater than 1.5 million acre-feet (25,000 cfs average monthly flow).

The Center and the Technical Work Group have developed several beach/habitat-building flow scenarios (e.g., in terms of flow duration, magnitude, and load following alternatives [Melis et al., 1998]), as well as a "resource criteria" procedure to ensure systematic and timely responses to hydrologic conditions, impact assessment, and compliance requirements in the event that a hydrologic triggering event does occur (Ralston, Winfree and Gold, 1998). The Center and the Technical Work Groups are continuing to examine hydrologic forecasting models and likely frequencies of beach/habitat-building flow events.

The Adaptive Management Work Group is consistently described in Center and Adaptive Management Program documents as being composed of various "stakeholder" groups. This term is more generally used in resource management to refer to potentially affected parties. Given the Grand Canyon's national and international significance, the full range of potential "stakeholders" is very large. As the term is used in the Adaptive Management Program and throughout this report, however, it generally refers to a more specific group of federal

TABLE 1.1 Operating Limits of the Secretary's Record of Decision*

Minimum releases:	8,000 cfs between 7 a.m. and 7 p.m. 5,000 cfs at night
Maximum releases:	25,000 cfs (exceeded during beach/ habitat-building flows)
Allowable daily fluctuations:	5,000, 6,000, or 8,000 cfs
	5,000 cfs: Daily fluctuation limit for monthly release volumes less than 600,000 acre-feet
	6,000 cfs: Daily fluctuation limit for monthly release volumes of 600,000–800,000 acre-feet
	8,000 cfs: Daily fluctuation limit for monthly release volumes of over 800,000 acre-feet
	Ramp rates[1]: 4,000 cfs/hour up 1,500 cfs/hour down

*Subject to emergency exception criteria for emergency releases and continuing discussion of hydropower "regulation" fluctuations.

[1]Ramp rates indicate the limits at which discharge through the dam can be increased ("up") and decreased ("down").

resource management agencies; Indian tribes; Colorado River Basin state representatives; and nongovernmental organizations representing environmental, recreation, and hydropower interests. The current Adaptive Management Work Group (see Appendix B) includes twelve "cooperating agencies" (including six tribal groups), representatives from the seven basin states, and two representatives each from three nongovernmental groups (environmental, recreational, and federal power purchasers). The Adaptive Management Work Group is the primary stakeholder group in the Adaptive Management Program.

The Technical Work Group (Appendix C) includes representatives from the AMWG's cooperating agencies and other members. Some of these agencies are primarily management organizations, such as the Bureau of Reclamation and the National Park Service; others are primarily science organizations, such as the U.S. Geological Survey; and others are nongovernmental organizations. The aim of the Technical Work Group is "To articulate to the GCMRC the science and information needs expressed in the objectives defined by the AMWG, and to assist in recommending science priorities" (Center, 1997, p. 28). Given their close relations to the Adaptive Management Work Group and their interests in Grand Canyon monitoring and research, Technical Work Group members may be considered "stakeholders" serving as technical representatives in the Adaptive Management Program.

When the Glen Canyon Dam Environmental Impact Statement was completed in March 1995, a "Transition Work Group" was created to help effect a transition from the Glen Canyon Environmental Studies Phase II to the Adaptive Management Program. It drafted guidelines, protocols, and administrative plans for the Center and worked on management objectives and information needs for the future. The Center joined the Transition Work Group in November 1995 to begin formulating the Center's Strategic Plan.

The Center released a final version of the Strategic Plan for 1997–2002 (Center, 1997) on May 1, 1997 (see http://www.gcmrc.gov). The plan was quickly approved by the Adaptive Management Work Group at its first meeting in September 1997. The National Research Council was asked to review the plans in January 1998. A committee was convened and began its work in May 1998.

CHARGE TO THE COMMITTEE

The National Research Council committee was charged to address two main questions and five related questions regarding the Long-Term Strategic Plan and fiscal year 1999 Annual Plans:

1. Review the Long-Term Plan using interdisciplinary input to determine if the current Grand Canyon Monitoring and Research Center plan will be effective in meeting requirements specified in the Grand Canyon Protection Act and the Glen Canyon Dam Environmental Impact Statement and Record of Decision. At least three objectives [questions]

must be evaluated to determine if the above requirements are met:

Objective 1: Does the Long-Term Plan respond to the new adaptive management process called for by the Act and the Glen Canyon Dam Environmental Impact Statement? That is, is the Grand Canyon Monitoring and Research Center functioning effectively in the Adaptive Management Program, especially regarding incorporation of all stakeholder objectives and information needs in the planning process?

Objective 2: Does the Long-Term Plan incorporate past research knowledge in developing new monitoring and research directions?

Objective 3: Has the Grand Canyon Monitoring and Research Center appropriately addressed past reviews of Glen Canyon Environmental Studies programs in formulating new research directions?

2. Characterize weaknesses of the Long-Term Plan and recommend short and long-term science elements to the Grand Canyon Monitoring and Research Center to address identified weaknesses. Two objectives [questions] must be addressed to respond to this goal:

Objective 1: What weaknesses exist in the Long-Term Plan and how do these weaknesses affect the potential effectiveness of the overall science program?

Objective 2: What changes can be made to the Long-Term Plan to overcome defined weaknesses and/or enhance the Long-Term Plan to meet its defined mission? What specific science elements (programs) are necessary to correct specific plan weaknesses?

The strategic plans encompass the Center's policy mandate, its perspective on adaptive management, its monitoring and research programs, and its budget. An assessment of whether the Center is functioning effectively in the Adaptive Management Program requires analysis of organizational and staffing issues. Although it was not charged to do so, this committee identified strengths as well as weaknesses of the Strategic Plan to provide a balanced review and to recognize important accomplishments of the Center and the Adaptive Management Program. In some cases, the committee identified specific science elements for improving Center programs. In other cases, guidance is offered at a general level. In yet other instances, solutions were not immediately clear and will have to be addressed by the Center and Adaptive Management Program stakeholder groups over the long term and with use of the Strategic Plan.

This review builds on previous National Research Council reviews of the Glen Canyon Environmental Studies (Table 1.2). National Research Council reviews of Colorado River management, more broadly defined, date back to *Water and Choice in the Colorado River Basin: An Example of Alternatives in Water Management* (NAS, 1968), which noted changing attitudes toward dams but did not examine dam-operations alternatives in detail. It anticipated debates about the range of alternatives that, in the broader public forum but not in the Adaptive Management Program, has included draining Lake Powell (U.S. Congress, 1997). Earlier scientific reviews of Colorado River development identified issues related to flooding, sediment transport, wildlife, and recreational effects of dam construction on the Colorado's mainstem (President's Water Resources Policy Commission, 1950; U.S. Bureau of Reclamation, 1950; U.S. Geological Survey, 1925; U.S. National Park Service, 1946). A 1946 report, *The Colorado River: A Natural Menace Becomes a National Resource*, included the following comment from the U.S. Fish and Wildlife Service: "The methods of reservoir operation, therefore, will be the determining factors in mitigation of damages and possible creation of benefits" (U.S. Department of the Interior, 1946, p. 252).

Fifty years later, a National Research Council (1996a) committee reviewed what had been learned in the Glen Canyon Environmental Studies, the most comprehensive investigation of dam-operation effects attempted to date. The review noted progress toward an ecosystem framework, external peer review, and administrative organization. Since then, a major controlled flood has been released, the Center has been established, and the Adaptive Management Program has been launched. As part of that Program, the Center prepared a Strategic Plan and requested a review by the National Research Council. In light of the past experience with the Glen Canyon Environmental Studies, it was not surprising that the Strategic Plan and National Research Council review were further complicated by unfolding events.

COMPLICATIONS WITH THE STRATEGIC PLAN

The 1997 Strategic Plan was adopted with an informal understanding that unresolved issues would be addressed after the Center was established. These unresolved issues included "potential new management objectives and information needs, and a proposed Lake Powell program" (Technical Work Group Minutes, 1997, p. 18). A conceptual model and research syntheses were also intended to produce "increased

TABLE 1.2 NRC Reports on the Colorado River

1987	River and Dam Management (National Academy Press).
1988	"Supplementary Report to River and Dam Management."
1988	"Letter report to the Honorable Donald Paul Hodel."
1991	Colorado River and Dam Ecology. Symposium proceedings (National Academy Press).
1991	"Review of the Draft Integrated Research Plan for the Glen Canyon Environmental Studies, Phase II."
1991	"Letter report to Commissioner Dennis Underwood."
1991	"Letter report to the Honorable Manuel Lujan."
1992	"Letter report to Michael Roluti...on May 1992 draft report 'Power system impacts of potential changes in Glen Canyon power plant operation.' "
1992	"Letter Report to David L. Wegner...assessing proposed GCES studies related to economics, hydropower production and dam operations."
1992	"Long-Term Monitoring Workshop for the Grand Canyon," position papers.
1993	"Letter report to Tim Randle...on January 1993 preliminary draft 'Operation of Glen Canyon Dam, Colorado River Storage Project, Arizona.' "
1994	Review of the Draft Environmental Impact Statement on Operation of Glen Canyon Dam.
1994	Review of the Draft Federal Long-Term Monitoring Plan for the Colorado River below Glen Canyon Dam.
1996	River Resource Management in the Grand Canyon (National Academy Press).

SOURCE: NRC (1996a, pp. 8–10).

knowledge to revise the Strategic Plan" (Center, 1997). The Center thus chose to revise the Strategic Plan soon after this National Research Council committee started its review. The revised plan was reviewed by the Technical Work Group in September and November 1998 and was to be approved by the Adaptive Management Work Group in January 1999. As these revisions presented a moving target, the National Research Council committee decided to assess both the current 1997 Strategic Plan and the 1998 draft Strategic Plan.

The situation became more complicated in December 1998, however, when the Technical Work Group decided it could not recommend the revised plan for adoption by the Adaptive Management Work Group. Some stakeholders expressed serious concerns about sections that dealt with policy, adaptive management, geographic scope, and Center administration. These issues had been postponed to get the Program off the ground, partly with the hope that they would be resolved within and through the Adaptive Management Program.

The adaptive management process yielded further complications when discussion of the Center's draft Strategic Plan revealed the lack of a strategic plan for the overall Adaptive Management Program. At the January 1999 Adaptive Management Work Group meeting, Chapters 1–3 of the Strategic Plan were reassigned to the Technical Work Group: "The TWG should focus on the Strategic Plan for the Adaptive Manage-ment Program first using the draft that was developed by the GCMRC [i.e., Center], and completing the final draft for review and approval by the next AMWG meeting" (http://130.118.161.89/amwg_new/quick-_updates.htm–1/20/99). The Technical Work Group subsequently divided this motion into two tasks, the first of which was to draft a "Guidance Document" of existing laws and policies defining the overall scope of the Program (in consultation with the U.S. Department of the Interior's Office of the Solicitor); the second task was to prepare a strategic plan for the entire Program.

These events indicate the salience and the complexity of strategic planning for monitoring, research, and adaptive management in Grand Canyon. Some events originate from criticisms of the Glen Canyon Environmental Studies, such as some stakeholder concerns about expanding scientific programs and increasing budgets. To date, there has been no detailed historical assessment of the Glen Canyon Environmental Studies and its bearing upon adaptive management and ecosystem science in the Grand Canyon. A 1990 National Research Council symposium on Colorado River Ecology and Dam Management included two brief chapters on the history of the Glen Canyon Environmental

Studies written by leading participants in that program (chapters by Wegner and Patten in NRC, 1991. cf. NRC, 1996a). The 1997 Strategic Plan contains a historical synopsis, but it does not analyze its implications for monitoring, research, or adaptive management. Other important aspects of Grand Canyon use and management have received historical attention (Lavender, 1985; Martin, 1990; Morehouse, 1996; Pyne, 1998; Schmidt et al., 1999a).

The Center should encourage professional historians to examine the record of scientific contributions to management of the Grand Canyon river ecosystem. Although this committee did not include a historian, it encourages archiving at the Center to facilitate historical analysis of what has and has not proven "adaptive" in the Grand Canyon ecosystem.

METHODS FOR REVIEWING THE STRATEGIC PLAN

The Center's "Strategic Plan" reviewed in this report has three components: (1) the 1997 Strategic Plan, which is still in force, (2) the 1998 draft revisions and debates about them, and (3) monitoring and research chapters of the 1998 draft plan that will form the basis for the Center's new plan (Figure 1.3). To evaluate these plans, the committee employed multiple sources of information and methods of review (Knaap and Kim, 1998; Shadish et al., 1991). The principal methods involved document analysis and discussions with Center staff. The Center provided copies of plans, Adaptive Management Program documents, requests for proposals, and copies of successful proposals. Minutes of Adaptive Management Work Group and Technical Work Group meetings were obtained from the Internet.

Committee members contacted individual Adaptive Management Work Group and Technical Work Group members for their views about Center plans and programs. Several committee members attended Adaptive Management Work Group meetings in Phoenix in July 1998, January 1999, and July 1999, which provided a deeper understanding of the Center's relations with other groups in the Adaptive Management Program. Two committee members participated in science trips in the Grand Canyon. Four participated in conceptual modeling workshops in October, November, and December 1998. One participated in the protocols evaluation program. Two members attended Technical Work Group meetings and ad hoc meetings in November 1998 and February 1999.

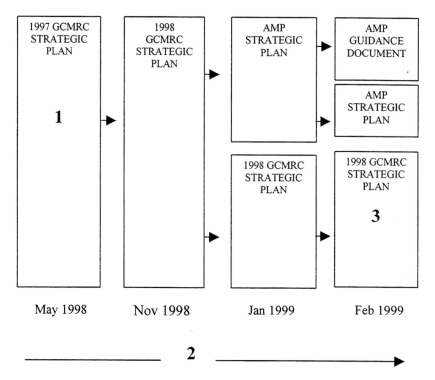

FIGURE 1.3 Evolution of the Long-Term Strategic Plan.

These activities led to a report that parallels the structure of the Center's Strategic Plan and speaks to questions posed in the committee's charge. Chapter 2 discusses the challenges of strategic planning. It examines the evolution of the Center's strategic plans and identifies general strengths and weaknesses, recognizing that the Center is a new organization, which calls for formative, rather than summary, evaluation (Rossi and Freeman, 1993). Nevertheless, we strive for a preliminary response to the question of whether the Strategic Plan will be effective in meeting requirements specified in the Grand Canyon Protection Act, Glen Canyon Dam Environmental Impact Statement, and the Record of Decision.

Chapter 3 examines the Center's evolving roles in the Adaptive Management Program. It asks whether a common understanding of adaptive management and the Center's roles in it have emerged, and discusses the implications of both common and pluralistic visions for Grand Canyon resources. It discusses the management objectives and information needs that guide the Center's research and monitoring program, and it evaluates the developing roles of independent review panels. In these ways it addresses the question, "Does the Long-Term Plan respond to the new Adaptive Management Process?"

Chapter 4 addresses the core of the Strategic Plan. It reviews the overarching framework for ecosystem science and monitoring. It assesses the Center's five main resource program areas: (1) physical resources, (2) biological resources, (3) cultural resources, (4) socio-economic resources, and (5) information technology. In each case, it asks how well the program areas incorporate previous reviews of the Glen Canyon Environmental Studies and previous research knowledge, identifying strengths, weaknesses, and alternatives.

Chapter 5 turns to organizational resources, including budget, staff, and administration. Many debates among stakeholders and scientists in the Adaptive Management Program involve organizational issues. These issues thus have an important bearing on the Center's Strategic Plan and on whether the Center is likely to fulfill the requirements of the Grand Canyon Protection Act, the Glen Canyon Dam Environmental Impact Statement, and the Record of Decision which is thus the opening and concluding question for this review. Chapter 6 draws together the report's main findings and recommendations.

2

The Center's Long-Term Strategic Plan

The Center requires a range of strategies to fulfill its mandate. Responding to management objectives and information needs demands flexibility. Long-term monitoring, by contrast, requires stable strategies for measurement and data management, while research programs should encourage innovation and creativity (Holling, 1998). Effective communication of scientific results depends on sound strategies for social learning and group decision-making (Gunderson et al., 1995; Lee, 1993; Walters, 1997).

These different types of strategies also require different types of evaluation (Mastrop and Faludi, 1997; Mintzberg, 1990; Mintzberg and Waters, 1998; Mintzberg et al., 1998; Westley, 1995). Monitoring plans lend themselves to formal evaluation of goals, protocols, and outcomes. Research programs require peer review. Evaluating the contributions of science to adaptive management is even more complex and may involve participant observation and surveys, as well as broad interdisciplinary reviews.

An overarching strategic challenge for the Center is thus to articulate the relationships among all these Program elements. Coordination is especially important in complex ecosystem-level programs. A strategic plan enables participants to envision common goals and agree on ways and means to achieve them. Where objectives compete with one another, it guides the collection of information on the likely consequences of alternative courses of action. It helps ensure that participants understand program goals, how those goals are to be achieved, and how unexpected events and surprises can be addressed. In an adaptive management program, a strategic plan should articulate management goals and alternatives, providing a framework for interpreting experimental outcomes and evaluating trade-offs and compromises among alternative management actions.

This chapter reviews the general strengths and weaknesses of the Center's Long-Term strategic plans to address whether the Plan will be effective in meeting requirements specified in the Grand Canyon Protection Act, the Glen Canyon Dam Environmental Impact Statement, and the Record of Decision. It draws upon knowledge in the field of strategic management (Mastrop and Faludi, 1997; Mintzberg, 1990; Segal-Horn, 1998) and upon committee members' views.

THE CENTER'S STRATEGIC PLANS

The Center has prepared two long-term strategic plans. The first was written in May 1997 and adopted later that year. The second was drafted in November 1998 but not adopted. Debates about the revised Plan arose in part from policy issues that had been delayed until the Adaptive Management Program was established. Adoption of the 1997 Strategic Plan and rejection of the 1998 Plan may be indicative of changes within the Adaptive Management Program, as well as unresolved issues in the strategic plans.

The 1997 Strategic Plan aimed to "implement the adaptive management and ecosystem science approaches" called for in the Grand Canyon Protection Act, the Glen Canyon Dam Environmental Impact Statement, and the Record of Decision. It also proposed to "build upon the rich history of monitoring and research investigations developed by the Bureau of Reclamation and other organizations" (Center, 1997). The 1998 Strategic Plan stated some related aims: (1) to describe Center programs, (2) to develop the programs cooperatively with the Adaptive Management Work Group, and (3) to provide a guidance document for annual plans. Both documents included introductory chapters on institutions and the adaptive management paradigm (Table 2.1). They mentioned the Grand Canyon Protection Act, Glen Canyon Dam Environmental Impact Statement, and the Record of Decision that mandate the Adaptive Management Program, as well as the "Law of the River" and other laws that constrain the Program (cf. Harris, 1998, for a broader inventory). The 1997 Strategic

TABLE 2.1 Structure of the Center's Strategic Plans

1997 Strategic Plan	1998 Strategic Plan	1999 Final Plans
CH 1: History of Monitoring and Research in the Grand Canyon (6 pp)	CH 1: Introduction— Purposes and Background (6 pp)	CH 1: Introduction
CH 2: GCMRC Program Justification and Mission (3 pp)	CH 2: Glen Canyon Dam Adaptive Management Program (16 pp)	CH 2: Philosophy of Monitoring
CH 3: Science Programming within Adaptive Management (16 pp)	CH 3: Management Objectives and Information Needs (7 pp)	CH 3: Monitoring and Science Programs
CH 4: Strategic Research Planning under Revised Paradigm and Institutional Constraints (12 pp)	CH 4: Scientific Philosophy of Monitoring (11 pp)	CH 4: Schedule and Budget
CH 5: Defining Stakeholder Objectives and Management Information Needs (3 pp)	CH 5: Monitoring and Science Programs (85 pp)	
CH 6: Monitoring and Science Programs (72 pp)	CH 6: Schedule and Budget (3 pp)	
CH 7: Schedule and Budget (8 pp)	Literature Cited	
	Appendices	
Appendices		

Plan refers briefly to some aspects of federal trust responsibilities related to Indian tribes (cf. Tsosie, 1998, for broader treatment). Both documents address the Adaptive Management Program's geographic scope.

Chapters in the Strategic Plan on the Center's monitoring and research programs changed substantively from 1997 to 1998. The 1998 Strategic Plan gave less attention to the adaptive management paradigm and more to a "philosophy of monitoring," which indicated a growing emphasis on monitoring programs. The 1998 Strategic Plan combined cultural and socioeconomic programs under the heading of sociocultural resources in an effort to develop a broader and more integrated approach to assessing the social effects of dam operations. The 1998 Plan gave less attention to contingency planning than did the 1997 Plan, which seems unwise in light of the importance of preparing for "surprises" in adaptive management. Both the 1997 and 1998 Plans concluded with a brief chapter on schedule and budget. Neither plan included a discussion of staffing, management, or organizational strategies.

When the Adaptive Management Work Group removed Chapters 1–3 of the 1998 Plan—leaving only the chapters on philosophy of monitoring, resource programs, and schedule and budget—it raised both a potential problem and an opportunity. Separating science and adaptive management plans could increase problems of coordination and seems to run counter to the aim of coordinating science and policy. In the short term, however, separating the two sections presents an opportunity for stakeholder groups to clarify policy issues while the Center refines its monitoring and research programs. A draft outline of a "Guidance Document," prepared by the Technical Work Group in collaboration with the Office of the Solicitor, seems promising. It is planned to be more comprehensive and detailed than the Center's treatment of institutional issues (Technical Work Group, 1999; cf. Rogers [1998] on adaptive management programs).

In 1996, the National Research Council recommended that a planning group be established separate from the ecosystem study group to implement the Adaptive Management Program. The Transition Work Group (1995–1996) prepared plans separately from the Glen Canyon Environmental Studies. That separation of planning and study responsibilities dissolved during the Center's first two years, but may reassert itself with preparation of the Adaptive Management Program Guidance Document and the Adaptive Management Program Strategic Plan. When they are complete, however, the Guidance Document,

Adaptive Management Program Strategic Plan, and Center Strategic Plan must fit closely and interact well with one another.

STRENGTHS OF THE PLAN

Center scientists report that they have used the Strategic Plan to guide annual planning. Interviews with members of the Adaptive Management Work Group and Technical Work Group (the stakeholder groups) yielded a range of views of the Strategic Plan's utility. Some found it a useful reference and consulted it before meetings; others regarded it as overly expansive in scope and length; still others attached little importance to it. Aside from specific points of criticism, discussed later, there were no clear patterns of use and evaluation by different stakeholder groups.

The Center's efforts nonetheless have established the salience of strategic planning. Controversy over the 1998 Strategic Plan, while rooted in deeper unresolved policy issues, has had the positive effect of bringing those issues to the surface where they are now being addressed. Although the strategic plans rightly referred to policies that both enable and constrain the Center, the decision to move Chapters 1–3 to a Guidance Document would allow a more complete treatment of those chapters. Indeed, some branch of the Adaptive Management Program should undertake continuing institutional analysis. Previous National Research Council reviews of the Glen Canyon Environmental Studies included chapters on institutional issues that affect ecosystem monitoring and research, such as interagency relationships, policy changes, external reviews, and the roles of funding (Ingram et al., 1991; NRC, 1987, 1996a; cf. NRC, 1996b). Institutional and legal analyses are not explicitly incorporated in the Center's plans for adaptive management, socioeconomic research, or external review. The strategic plans were likewise on firm ground in attempting to situate monitoring and research within the broader context of adaptive management.

Strengths of the 1997–98 strategic plans to date are thus their roles in guiding annual plans, their attempts to coordinate science and adaptive management, and their salience in raising unresolved issues for discussion. Until recently, the entire responsibility for developing strategic plans fell on the Center. Although perhaps reasonable initially, a broader distribution of responsibility for preparation of strategic plans with the Technical Work Group would enable the Center to focus more on monitoring and research.

WEAKNESSES AND ALTERNATIVES

The strategic plans have five general weaknesses: (1) insufficient definition of the Center's key strategic priorities to be addressed in the next five years, (2) inadequate discussion of geographic scope, (3) neglect of medium- and long-term time scales, (4) insufficient attention to the public significance of monitoring and research, and (5) omission of organizational and resource issues (e.g., staff and coordination).

Defining Key Strategic Priorities

The plans describe proposed monitoring and research programs without appraising: (1) what has and has not been accomplished, (2) challenges that stand out for the next five years, and (3) how proposed programs would build upon accomplishments and address failures. The committee recognizes there are many ways to articulate planning strategies, but considers these three elements crucial (cf, Mintzberg, 1998; Westley, 1995).

The 1997 Strategic Plan summarized the legacy of the Glen Canyon Environmental Studies, but did not analyze it or use it to justify the proposed Plan. Similarly, the 1998 Plan did not discuss what has and has not been accomplished since the Center was founded in November 1996. Among other things, the transition from the Glen Canyon Environmental Studies to the Center was completed. Charters were written, staff hired, adaptive management meetings convened, protocols defined, research grants awarded, a conceptual model built, logistics centralized, and a split of Lake Powell monitoring responsibilities negotiated. Stakeholders reported that the current situation is more collegial and promising than earlier, although some perceive that aspects of the process sometimes threaten to break down.

At the same time, monitoring programs have developed slowly. Although informal reasons were offered to explain the lack of implementation, these reasons could be addressed more explicitly in the Strategic Plan. The resource programs follow different approaches that are not well coordinated. Although different approaches may be needed, a coordinating strategy is essential. Relations with stakeholder groups have proven time-consuming and sometimes difficult; program scope and responsibilities remain contested.

These accomplishments and problems define the Center's current situation. They indicate the challenges that must be addressed and suggest precedents, analogues, and alternatives for addressing them. The main challenge when the 1997 Plan was written was to establish the Center and define its programs. Now that the Center is established, what are the main challenges for the next five years? The 1998 Plan has elements of a "problem statement" in a section on current science needs and chapter on the philosophy of monitoring, but that chapter is more a list of factors to consider than strategic challenges to address.

Based on the committee's review of the strategic plans, the Center may wish to give greater attention to the following key challenges:

(1) implementing a long-term monitoring program, (2) clarifying the scientific basis of the existing adaptive management experiment, (3) coordinating monitoring and research in the resource program areas (and with related agency programs), (4) resuming socioeconomic research and decision analysis, (5) increasing the efficiency and effectiveness of Center participation in stakeholder processes, (6) implementing information management, and (7) contingency planning for environmental and policy surprises.

Although the Center may decide that other issues have even higher priorities, the point here is that it should identify the top strategic priorities for the next five years.

The aims and methods of strategy formulation are changing in ways that have a bearing upon the Center's plans. During the 1960s and 1970s, strategic planners emphasized optimization methods and measurable goals and outcomes as formal criteria for evaluating program performance. That formal approach may still be appropriate for monitoring programs where one wants to know whether wise choices have been made about what to measure, whether those measurements are accurate, complete, and systematically recorded, and whether a systematic plan has been established and followed to implement the monitoring program. By the 1980s, however, many formal strategic plans failed to materialize, or they constrained organizational changes necessary to improve performance. Similarly, comprehensive plans for water development in the Colorado River Basin fell behind changing public attitudes and demands. Greater emphasis has subsequently been placed on strategies for adapting, learning, positioning, and envisioning (Mastrop and Faludi, 1997; Mintzberg, 1998; Westley, 1995)—all of which resonate closely with adaptive environmental management as it is developing in the Grand Canyon.

Westley (1995) distinguished "planning," "visionary," and "learning" strategies, and stressed the importance of managing changes and cycles among these strategies. The Glen Canyon Environmental Studies had a visionary aspect that was later followed by greater emphasis on planning and learning through adaptive management. The overarching strategic challenge now is to coordinate these "planning," "visionary," and "learning" strategies (Mintzberg, 1998). The lack of clear coordination among the resource programs and adaptive management activities is evident in graphic representations in the strategic plans (Figures 2.1 and 2.2). These diagrams do not clearly depict the relationships among Program elements, processes, roles, and functions. Refining these graphic diagrams could help clarify a strategy for coordinating the Center's monitoring, research, and adaptive management roles. Other diagrams might focus on coordination of the five resource programs—physical resources, biological resources, cultural resources, socioeconomic resources, and information technology—that presently follow different outlines and approaches.

Geographic Scope

The geographic scope of the Adaptive Management Program has generated debate. Some stakeholders want sharper "sideboards" (boundaries), while others seek to include geographic linkages with upstream, downstream, and tributary processes.

The mandated focus of the Adaptive Management Program is on the effects of the Secretary's actions at Glen Canyon Dam on downstream resources. The Strategic Plan describes the Program's scope as the Colorado River ecosystem within Glen Canyon National Recreation Area

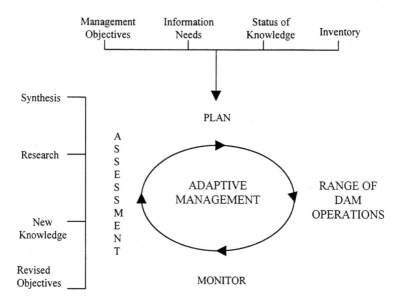

FIGURE 2.1   The Grand Canyon Monitoring and Research Center's approach to adaptive management. SOURCE: Center (1997).

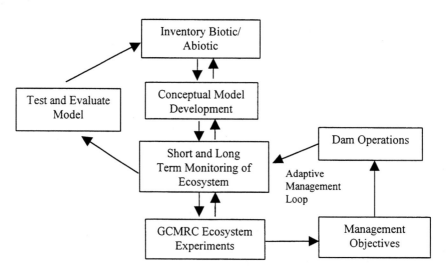

FIGURE 2.2   The Grand Canyon Monitoring and Research Center's approach to ecosystem and adaptive management. SOURCE: Center (1998).

and Grand Canyon National Park, which is defined as "the Colorado River mainstem corridor and interacting resources in associated riparian and terrace zones, located primarily from the forebay of Glen Canyon Dam to the western boundary of Grand Canyon National Park" (Center, 1998). The Program's lateral extent of the monitoring effort is defined by processes and conditions associated with dam discharges and river flows in connection with the Record of Decision. While this is defined "as the maximum regulated discharge and the inundated area for the annual pre-dam peak flows," it is also noted that "it is prudent in some areas of the Colorado River ecosystem to include elevations above the stage associated with flows of 100,000 cfs" (all quotes from Center, 1998). The Programmatic Agreement for cultural resources management has included surveys that extend laterally to the 256,000 cfs "old high water zone" flood stage.

The strategic plans left the door open for selected studies in Lake Powell, tributary watersheds, comparable river reaches elsewhere in the basin, and Lake Mead, if they are related to effects on downstream resources. That openness, along with its implications for budget and management, became a source of controversy.

The first test case involved water quality monitoring in Lake Powell. Negotiation of a five-year Lake Powell monitoring program led to an agreement known as the "Lake Powell split," which divided monitoring responsibilities into the following categories (see also Appendix G):

1. "White" — Adaptive Management Work Group Management objectives and information needs that relate to downstream (below Glen Canyon Dam) effects and include monitoring and research activities conducted downstream of Glen Canyon Dam. They are funded by the Adaptive Management Program budget, with the scope of work reviewed by the Adaptive Management Work Group and Technical Work Group.

2. "Gray" — Adaptive Management Work Group management objectives and information needs that relate to downstream effects, but include monitoring and research activities conducted upstream of Glen Canyon Dam. These are part of the Adaptive Management Program and use its procedures, they are funded by the U.S. Bureau of Reclamation or through the operation and maintenance budget or other sources, and the scope of work is developed by the Center and coordinated with the U.S. Bureau of Reclamation and other agencies.

3. "Black" — These are not directly related to downstream effects and include monitoring and research conducted upstream of the

Glen Canyon Dam. They are funded by the U.S. Bureau of Reclamation, other agencies, or other funding sources, and they are not formally a part of the Adaptive Management Program.

This classification indicates that the key issues are: Is there a physical connection between dam operations and downstream resources? Who conducts the research and monitoring and with what science protocols? Who pays for the research and monitoring? By addressing each of these issues, the three-category "Lake Powell split" moves beyond simple binary, but ecologically and institutionally flawed, distinctions between what is "inside or outside the box." It also represents a good example of adaptive management because it accommodates important resource linkages without losing geographic focus (U.S. Bureau of Reclamation, 1995). But it does not indicate how future decisions about geographic linkages will be made, or whether different criteria should be used for decisions about monitoring and research.

After reviewing these issues, the committee concluded that:

- **The Lake Powell split provides a useful model for addressing issues of geographic scope that will regularly arise in the future.**

Issues related to the boundaries of the Adaptive Management Program will surely recur. These issues include interactions with the old high-water zone and upslope areas, seeps, and springs; tributary inputs of water, sediment, organic matter, and biota; interactions between the Grand Canyon riverine and Lake Mead delta ecosystems; regional hydroclimatic linkages with dam operations and their joint resource effects; comparisons with other dam-operation experiments, species recovery programs, and analogous reaches in the Colorado River Basin; and interregional comparisons of adaptive management experiments. By anticipating issues that will surely recur and by designing ways to efficiently address them, the Program could build upon experience gained with the Lake Powell split and address relevant resource linkages without losing its focus on the river ecosystem.

The Strategic Plan could make better use of previous National Research Council reviews and recent geographic and ecosystem management research on boundaries. The 1987 National Research Council review described boundaries similar to those currently proposed as "unnecessarily and unreasonably restrictive" (NRC, 1987), while the 1996 review criticized the expansion of scope under Glen Canyon Environmental Studies Phase II as "unstable and expansive" (NRC, 1996a). These are the twin pitfalls of an overly narrow or broad geographic scope. The 1996 review also presented three criteria for defining geographic scope: management options, resources, and the ecosystem concept. The Lake Powell split succeeded by using all three criteria, but it also indicates that defining common stakeholder interests constitutes a fourth criterion.

The committee thus concluded that:

• **Rigid geographic boundaries will not serve the Adaptive Management Program well. After clearly defining the Program's geographic focus, decisions about geographic linkages must be made on a case-by-case basis, taking into account management options, ecosystem processes, funding sources, and common stakeholder interests.**

Individual stakeholder concerns extend in many geographic directions, but adaptive management may help identify common concerns associated with dam operations and downstream resources. Tribal reports commissioned by the Glen Canyon Environmental Studies, for example, raise important questions about linkages between the Grand Canyon river ecosystem and wider landscapes (Ferguson, 1998; Hart, 1995; Roberts et al., 1995; Stoffle et al., 1994). Although these views have not been taken up in planning documents, they may be more widely shared by stakeholder, public, and scientific groups than is presently recognized (Frederickson, 1996; Morehouse, 1996).

On a scientific level, the Center has collaborated effectively with National Oceanic and Atmospheric Administration scientists on the implications of the 1997 El Niño for dam operations and downstream resources (cf. Pulwarty and Redmond, 1997).

The committee concluded that:

• **The Strategic Plan could use the Center–National Oceanic and Atmospheric Administration collaboration as one model for research on regional geographic linkages that have effects on downstream resources, and for interagency coordination. Other models include co-financing of research on events and phenomena that, when combined with dam operations, have joint downstream effects relevant for adaptive management.**

Research on geographic boundaries and ecosystem management indicates that no boundary satisfies all functions or purposes (Forman, 1996; Keiter, 1994; Morehouse, 1996; Prescott, 1987). It is not likely that a single boundary rule will be appropriate in all policy or management contexts, and it would constitute poor science and management to allow a predetermined rule to rigidly constrain the scope of science and monitoring. Rather, a procedure is needed to decide individual issues on their merits and their relevance for understanding the effects of dam operations on the Grand Canyon ecosystem. Boundaries help focus a program, but they should be used to guide the manner in which resource linkages are investigated rather than preclude investigation. A procedure and criteria for managing boundary issues could help the Center move beyond "expand–contract" struggles to a more efficient and scientifically reasonable treatment of resource linkages that arise. In addition to whether or not the proposed research falls within the Adaptive Management Program, and who is to pay for it, the procedure should indicate who decides and how the decision is made. Local boundary issues that arise in research proposals might be delegated to the Center and peer review panels, while regional research and monitoring activities might involve higher levels of oversight and approval.

It is encouraging to note that the draft outline for the Technical Work Group's Guidance Document (Technical Work Group, 1999) refers to adaptive management in other parts of the Colorado River Basin and in other regions of North America. The Center should maintain an awareness of experiments in related and comparable basins to ensure a broad range of alternatives and lessons from past experiences when planning for Grand Canyon ecosystem management. This should include the Upper Colorado River Basin Recovery Implementation Program, the San Juan River Basin Recovery Implementation Program, and the Multispecies Recovery Program in the Lower Colorado River, since all fishes of interest in the Grand Canyon are under management both up- and downstream from Lake Powell in the Green, San Juan, and Colorado rivers. It should also include a strategy for drawing lessons from adaptive management and dam operations in other regions, such as the Columbia River Basin, the Everglades, and the Upper Mississippi River Basin (Gunderson et al., 1995; Independent Scientific Group, 1996; Independent Scientific Group, 1996; Lee, 1993; NRC, 1996b; Sit and Taylor, 1998; Volkman, 1997; Walters, 1997; Walters et al., 1992). Individual Center scientists currently follow these other experiments, but there is nothing in their job descriptions, mission statement, or budget to sustain those personal commitments.

The committee concluded that:

• **The Strategic Plan should indicate how the Center would draw insights from adaptive management in other regions, especially those involving water resources management.**

A large program of comparative research is not envisioned, but rather a creative strategy and modest resources for drawing practical lessons from related experiments. Leading experts spoke and led the second Adaptive Management Work Group meeting in January 1997, and also lead the conceptual modeling project. These are good examples of communication that helps maintain a creative flow of ideas and avoids common pitfalls in adaptive management.

Medium and Long Time Scales

The Strategic Plan spans five years, which may be "long" for administrative purposes, but it is too short for ecosystem management. The 1998 Plan lists time scales from hourly to interannual and "pre-dam versus post-dam time periods." Although this last time scale is longer than the strategic planning period, the Plan does not explicitly discuss decadal or multidecadal time scales, nor does it address the "perpetuity" for which Grand Canyon National Park was established. These medium and long time scales are relevant and essential for planning, ecosystem monitoring, and adaptive management.

The multidecadal life span and population dynamics of fish species such as humpback chub and razorback and flannelmouth suckers, for example, influence the design of monitoring programs today. As Lee (1993, p. 63) states, "The time scale of adaptive management is the biological generation rather than the business cycle, the electoral term of office, or the budget process." Monitoring of long-lived species and decadal ecosystem processes entails decades of data collection and design of experiments of similar duration. Social changes over the same time scales are analyzed in histories of western water management but rarely considered in the design of monitoring programs (Lee, 1993; NRC, 1968; Pisani, 1992). To understand how changes in downstream resources are experienced and evaluated, those social effects must be monitored and analyzed.

Major institutions also change on decadal time scales. The 1998 Strategic Plan was criticized for straying into policy issues related to the Grand Canyon Protection Act, the Glen Canyon Dam Environmental Impact Statement, the Record of Decision, and the Law of the River. But these issues and changes bear directly on Center programs. During the course of this review, for example, the U.S. National Park Service considered wilderness designation for as much as 94 percent of Grand Canyon National Park, which would affect the conduct and costs of monitoring and research; biological opinions for endangered species were reviewed; and stakeholders debated the relative importance of sections 1802, 1804, and 1805 of the Grand Canyon Protection Act.

Section 1804 of the Act focuses on "the Colorado River downstream of Glen Canyon Dam." Section 1802 refers more broadly to the "values for which Grand Canyon National Park and Glen Canyon National Recreation Area were established, including, but not limited to natural and cultural resources and visitor use," and section 1805b to the "effect of the Secretary's actions on the...resources of Grand Canyon National Park and Glen Canyon National Recreation Area." In the short-term (inter-annual), these might be regarded as competing principles, but a longer-term, decadal perspective would reveal how they are logically related to one another (see Figure 2.3, and see discussion in Chapter 3).

Similarly, a historical perspective on the Law of the River indicates that major changes have occurred on a decadal frequency during the twentieth century (e.g., Harris, 1998). To design a long-term monitoring system without considering the likely occurrence and uncertain implications of such social and institutional changes and trends could reduce the Program's robustness and flexibility.

The committee concluded that:

• **The Strategic Plan should relate five-year planning to multidecadal planning, ecosystem monitoring, and adaptive management.**

Public Significance of Center Programs

The Strategic Plan should consider the broader public context and

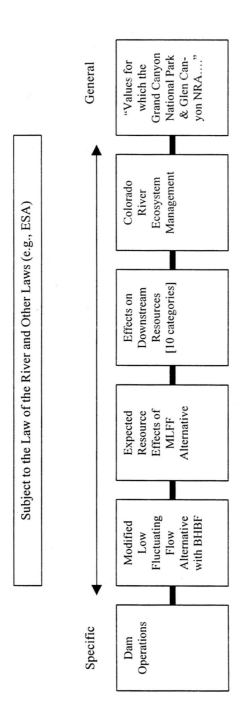

FIGURE 2.3 Institutional guides for monitoring and research.

significance of Grand Canyon monitoring and research.   Official stakeholders encompass a complex array of local to national interests, including federal agencies, Indian tribes, basin states, power consumers, and nongovernmental organizations. Public interests in "science" itself are not explicitly represented. The growing emphasis on science in adaptive management involves a broader, nationwide public movement to address resource management problems as science-policy experiments (Lewis, 1994; Tarlock, 1996).

The potential significance of these science-policy experiments includes and extends beyond the concerns of officially recognized stakeholders. In addition to its status as a national park held in trust for the citizenry of the United States, the Grand Canyon is a United Nations Educational, Scientific, and Cultural Organization (UNESCO) World Heritage Site with international and global significance. At every level of adaptive management, from ad hoc committees to the Secretary's actions, an aim of the scientific monitoring and research programs is to clarify and secure the "common interest" in the Grand Canyon ecosystem (Brunner and Clark, 1997).

The Center's strategic plans give little attention to public interest in science-based approaches to resource management. They are certainly open to public involvement, outreach, and education, but they have not explicitly made plans or budgets to respond to or encourage these activities. Although this is not surprising for a new program, a long-term strategic plan should anticipate the prospect of expanding public interest and involvement. An example is underway in the Upper Mississippi River Basin, where a program of "citizen science" (i.e., public education and museum activities) is envisioned to lay the foundation for broad social decision-making in future decades (S. Light, Minnesota Department of Natural Resources, personal communication, 1998).

The committee concluded that:

• **The Strategic Plan should explicitly recognize and speak to public interests in Grand Canyon monitoring and research and should anticipate programs of public education, outreach, and involvement.**

Over the long term and therefore in its Strategic Plan, the Center should also strive to use monitoring and research to clarify common interests in downstream resources.   While the Adaptive Management Work Group is responsible for articulating the common interest in management of Glen Canyon Dam and the Grand Canyon ecosystem, the Center is responsible for stewarding emerging public interest in science-based approaches to dam and ecosystem management.

Organizational Resources as Strategic Planning Issues

The strategic plans include brief chapters on "Schedule and Budget," but do not include discussions of the staffing, administration, or budget needed to achieve the plans' goals. These omissions contribute to ambiguity in how plans will be implemented and to continuing concerns about budget increases, which results in uncertainties for programs and personnel. The lack of sufficient cost and administrative information was a criticism voiced in a National Research Council review of the initial draft long-term monitoring plan (NRC, 1994).

The inclusion of anticipated resource needs for the Center in its Strategic Plan is very important. The 1997 Plan was more explicit on this point than the 1998 Plan. Neither of the strategic plans discusses staffing issues and needs, which are critical issues for both the Center and Adaptive Management Program. Questions of *what* the Center should do are inextricably linked with questions of *who* should do it (and how many staff are necessary) and should therefore be addressed. Some attention should also be paid to securing longer-term funding and contingency planning to eliminate the negative impacts of uncertain funding cycles on new and continuing science programs. For example, in the fiscal year 1999 budget cycle, the Upper Colorado River Recovery Implementation Program sought appropriations of $46 million to be spent over the next five-year period. At the same time, contingency planning is needed to respond to environmental and policy surprises (e.g., tributary flood events, and the proposed wilderness designation for Grand Canyon National Park).

SUMMARY

The question of whether the Strategic Plan will be effective in meeting the requirements specified in the Grand Canyon Protection Act, the Glen Canyon Dam Environmental Impact Statement, and the Record of Decision can be addressed in a preliminary way from the evidence above. The strategic plans indicate that Center scientists have a keen appreciation of these requirements. They cite relevant policy documents and grasp their logic, spirit, and limits. The strategic plans have been designed to fulfill those requirements, and the committee concluded that they have a good chance of success.

How *well* they meet those requirements, however, depends upon how clearly the plans define the Center's most pressing challenges for the next five years. This depends upon a wise combination of focus and flexibility when addressing issues of geographic scope, building upon previous experience when possible. It requires a broader view of time scales and the emerging public interest in scientific monitoring and research. It depends upon a pragmatic strategy for marshalling organizational resources in a turbulent environmental and policy context. Finally, and perhaps most important, it depends upon an innovative strategy for assuring independent scientific inquiry in a program of adaptive management.

3

The Adaptive Management Program

INTRODUCTION

This chapter assesses the Strategic Plan's responsiveness to the Adaptive Management Program. Although it focuses on adaptive management as defined in the Strategic Plan and developed in the context of Glen Canyon Dam operations and the Grand Canyon ecosystem, it begins with a brief overview of the emerging field of adaptive management. It then turns to strengths and weaknesses of the Strategic Plan's sections on adaptive management, the Plan's responsiveness to the Adaptive Management Program, and the roles of independent review in adaptive management.

In general, the Strategic Plan reflects the Center's efforts to respond to the new Program, in part by drawing upon general concepts and methods of adaptive management. However, the Plan also has some important weaknesses, as does the larger Adaptive Management Program, that impinge upon the Center's ability to fulfill its scientific monitoring, research, and communication responsibilities. When appraising weaknesses, we focus on the Center's evolving roles within the Program. We ask whether there is a common vision for Grand Canyon resources and whether the core adaptive management experiment has been clearly defined, communicated, and initiated. Because stakeholder-defined management objectives and information needs are intended to guide the Center's monitoring and research programs and provide measurable standards for evaluating adaptive management, we examine the current list of management objectives and information needs and ask whether they are likcly to fulfill these roles. As management experiments are conducted, a basis will be needed to evaluate their results and formulate recommendations to the Secretary of the Interior. We consider the potential roles of decision support methods in the complex trade-off analyses stakeholders will make when recommending future dam-operation experiments and adjustments. Because independent scientific review is a key component of adaptive management, we conclude with a discussion of independent review panels.

THE EMERGING FIELD OF ADAPTIVE MANAGEMENT

Adaptive management has received increasing attention and application in recent decades to problems of regional ecosystem management (for recent reviews, see Gunderson et al., 1995; Lee, 1993; and Walters, 1997). It arose from concerns that conventional resource management approaches have not adequately incorporated principles of ecosystem science (e.g., those related to ecosystem dynamics, disturbance regimes, and scientific uncertainty). It argues that these deficiencies tend to increase vulnerability to ecological "surprises" (e.g., extreme geophysical events, exotic species invasions, and dramatic species population changes) and decrease ecosystem resilience (i.e., the rate of recovery from disturbance). It further asserts that some problems and processes encountered in large-scale ecosystems and complex resource management regimes can only be understood through management experiments—sometimes referred to as "learning by doing" (Walters and Holling, 1990). Conventional efforts to address these complex problems seem increasingly bound up in policy "gridlock" among stakeholder, management, and scientific groups.

Adaptive management was envisioned as a new paradigm for addressing this suite of ecosystem science and ecosystem management problems through a dynamic interplay of science, management, and policy. Its core concepts and methods coalesced in the 1970s at the International Institute for Applied Systems Analysis in Laxenburg, Austria (Holling, 1978). Although those concepts and methods are still evolving and are applied in various ways, one useful working definition states that: "adaptive management is a systematic process for continually improving management policies and practices by learning from the outcomes of operational programs. Its most effective form—"active" adaptive management—employs management programs that are designed to experimentally compare selected policies or practices, by evaluating alternative hypotheses about the system being managed" (Nyberg, 1998, p. 2).

The key components of this and other working definitions include: (1) commitment to ongoing management adjustments based, in part, upon scientific experimentation, (2) shift from "trial and error" to formal experimentation with management actions and alternatives, (3) shift from fragmented scientific investigations to integrated ecosystem science, (4) explicit attention to scientific uncertainties in ecosystem processes and effects of management alternatives, (5) formal experimental design and hypothesis-testing to reduce those uncertainties and help guide management adjustments, (6) careful monitoring of ecological and social effects and of responses to management operations, (7) analysis of experimental outcomes in ways that guide future management decisions, and (8) close collaboration among stakeholders, managers, and scientists in all phases of these processes.

These concepts are not entirely new. As Lee (1993) indicates, they have close affinity with the pragmatic tradition in philosophy and public policy that developed during the twentieth century (e.g., Dewey, 1938; Lindblom, 1959; Wescoat, 1992). They seek to refine and integrate, as well as move beyond, established practices in scientific experimentation and resource management. For example, one important refinement underway applies Bayesian statistical methods to the design of adaptive management experiments (Sit and Taylor, 1998).

Adaptive management has been tested in various resource management contexts in North America and elsewhere. Early applications occurred in forest and fisheries sectors in the Pacific Northwest region of Canada and the United States (e.g., Lee, 1993; National Research Council, 1996b; Taylor et al., 1997). Other important water resources applications are underway in the Everglades, the Upper Mississippi River Basin, and in California's Bay-Delta ecosystem (Adaptive Environmental Assessment Steering Committee and Modeling Team, 1997; CMARP Steering Committee, 1988; Harwell, 1998; Independent Scientific Group, 1996; NRC, 1996b; Strategic Plan Core Team for the CALFED Bay-Delta Program, 1998; Volkman and McConnaha, 1993; Walters et al., 1992). Related ecosystem management activities in the Colorado River Basin include the Upper Colorado River Endangered Fishes Recovery Implementation Program, the San Juan Recovery Implementation Program, and the Multispecies Conservation Program in the Lower Colorado River Basin (Pontius, 1997).

Each of these programs seeks to balance resource use with ecosystem science, economic values, and public interests in ecosystem services. Some of them may have relevance, through formal comparison or analogy, with the Adaptive Management Program. For example, the Columbia River Basin program has established an effective Independent Scientific Advisory Board and Independent Economics Analysis Board (J. Volkman, Northwest Power Planning Council, personal communication, 1998), neither of which yet exists for the Grand Canyon. The Independent Scientific Group in the Columbia River Basin produced a vision statement, the "normative river concept," which may have relevance for developing a vision in the Grand Canyon. In light of scientific uncertainties and conflicting scientific evidence regarding Snake River chinook salmon restoration, scientists in the Columbia Basin convened a "Weight of Evidence Workshop" that may have relevance when Grand Canyon monitoring and research results need to be analyzed and interpreted.

At the same time, the practice of adaptive management is unique to each ecosystem. Programs are structured in different ways to address these unique features. Society has not yet perfected the social, economic, and institutional components of adaptive management needed in specific contexts (Gunderson et al., 1995; Holling, 1978; Lee, 1993; Walters, 1986, 1997). For all of the potentially useful points of comparison with other adaptive management programs, the Grand Canyon is unique in many ways. In addition to its singular physiographic landscape and ecological characteristics, it is situated in the heart of the Colorado River water management system. It has one of the most complex and contested organizational contexts for water resources management in the world, as evidenced by the array of stakeholders and managers engaged in the Adaptive Management Program. This situation calls for innovation as well as creative application of adaptive management concepts and methods to Glen Canyon Dam and the Grand Canyon ecosystem. With this overview of adaptive management in mind, we turn to the evolving application in the Grand Canyon.

ADAPTIVE MANAGEMENT OF GLEN CANYON DAM
AND THE COLORADO RIVER ECOSYSTEM

Adaptive management is a central theme and organizing framework in the Strategic Plan. The 1997 Plan states that "Adaptive management begins with a set of management objectives and involves a feedback loop between the management action and the effect of that action on the system. It is an iterative process, based on a scientific paradigm that treats management actions as experiments subject to modification, rather than as fixed and final rulings, and uses them to develop an enhanced scientific understanding about whether or not and how the ecosystem responds to specific management actions" (Center, 1997, p. 30). The Plan briefly discusses the role of dialogue among managers, stakeholders, and scientists; scientific "reality-testing" of management objectives; monitoring and experimental design of adaptive policies; and the close relationship between adaptive management and ecosystem management. Many of these aspects of the adaptive management experiment have not yet been formulated as testable hypotheses (e.g., about the types of dialogue that lead toward or away from adaptive policies; and the types of monitoring evidence that would lead to new management experiments and recommendations).

Adaptive management encompasses dam-operation experiments (such as controlled floods and daily flow regimes) hypothesized to achieve downstream ecosystem benefits; monitoring the effects of those experiments; research to explain those effects; design of new experiments to more fully achieve ecosystem benefits; and stakeholder-guided management experiments to weigh monitoring and research results when recommending dam-operation experiments and adjustments to the Secretary of the Interior. Adaptive management is thus a science experiment, a policy experiment, and a science-policy experiment. As will be discussed below, the hypotheses in these experiments have not always been clearly defined and formally tested.

The Glen Canyon Dam Environmental Impact Statement included adaptive management as a common element for all alternatives, and the Record of Decision subsequently mandated its implementation. Adaptive management builds upon the efforts of the Glen Canyon Environmental Studies, which explored several lines of inquiry to develop an ecosystem framework that assisted the search for, and evaluation of, dam-operation alternatives (U.S. Bureau of Reclamation, 1995). The Adaptive Management Program strives for this approach by designing monitoring and research programs to provide advice to the Secretary of the Interior about dam operations that preserve and enhance downstream resources.

The Glen Canyon Dam Environmental Impact Statement defined adaptive management as a process "whereby the effects of dam operations on downstream resources would be assessed and the results of those resource assessments would form the basis for future modifications of dam operations" (U.S. Bureau of Reclamation, 1995). It envisioned that the

Adaptive Management Program would provide an annual report to Congress and to the governors of the Colorado River Basin states. It also listed a set of principles and goals related to monitoring and research, coordination and communication, public participation, effective use of scientific information, and conflict resolution. It described how the Center would assist the Secretary's designee and the Adaptive Management Work Group by developing annual monitoring and research plans, and by managing and coordinating adaptive management research programs and the data collected in these programs.

Given the Center's multiple roles, it is impossible to evaluate the Strategic Plan without examining its relations within the broader Adaptive Management Program. Similarly, because adaptive management is the shared aim of these organizations, it is important to assess the collective understanding of adaptive management and how that influences the Center's scientific programs. Set in this context, the Center and the Adaptive Management Program are participants in a large-scale science-policy experiment involving environmental management constructs that remain unproven and not well understood.

The prominence of the Grand Canyon National Park and the Glen Canyon and Lake Mead National Recreation Areas elevate the Adaptive Management Program to a national scale of importance, as indicated by passage of the Grand Canyon Protection Act in 1992. This act focused on protecting the river corridor in Grand Canyon from adverse impacts of Glen Canyon Dam operations. The Adaptive Management Program is in many ways more delimited than other adaptive management programs, involving only a segment of the river within relatively well-defined geographic boundaries. The management actions involve operations of Glen Canyon Dam, which reduces the practical set of alternatives under consideration. Unlike some national-level efforts, however, the Adaptive Management Program is not dominated by a single resource issue (e.g., salmon recovery). Restoration of endangered species and impending loss of biodiversity are often the dominant issues in these other programs. Moreover, the Grand Canyon is one of the only adaptive management programs to have its own monitoring and research center.

To date, the Adaptive Management Program has not produced a scientific and stakeholder-based consensus regarding the desired state of the ecosystem (Marzolf et al., 1998; Schmidt et al., 1998). Before the Adaptive Management Program can measure its success, it must first develop a clear statement of what it is trying to accomplish. The controversy over the first three chapters of the 1998 Strategic Plan indicates the need for broader understanding and acceptance among stakeholders of tenets of adaptive management as they apply in the Grand Canyon. It argues for continuing efforts to clarify the definition, aims, and methods of the Adaptive Management Program. Until a common definition of adaptive management is articulated and accepted in the Program, the Center will lack the guidance necessary to perform its function within the Program or to effectively revise its Strategic Plan.

The Center's roles in the Program should be founded on: (1) management objectives and information needs identified by stakeholders, (2) ecosystem science to guide monitoring, explain observations, and add neglected information needs, and (3) scientific and stakeholder communication to facilitate "social learning" based upon the knowledge gained from monitoring and research. Without the first, research may wander from its goal of understanding the effects of dam-operation alternatives on downstream resources. Without the second, management objectives may lack an adequate foundation in underlying ecosystem processes. If management objectives and information needs are not integrated within an ecosystem science approach, they are unlikely to anticipate possible "surprises" in ecosystem responses. Without the third, monitoring and research results may go unused, and the learning necessary to refine and revise management objectives may not occur (Parson and Clark, 1995). A well-defined strategic plan would indicate how monitoring and research programs would build in a balanced way upon these three points.

STRENGTHS OF THE STRATEGIC PLANS

From its inception, the Center has performed a valuable service by articulating the aims, concepts, and methods of adaptive management. Although other Adaptive Management Program documents (e.g., charters and operating rules) describe the structure and procedures of adaptive management, the Center's strategic plans contain the most detailed discussion of the Program's philosophy and implementation in the Grand Canyon to date. Introductory chapters remind participants of the initiating roles of stakeholders' management objectives, feedback provided by scientists through ecosystem monitoring and research (including identification of key information needs), and consequent adjustments in both management and science. The strategic plans include major chapters on stakeholders' management objectives and information needs. The Center has helped facilitate their listing and prioritization. The plans also attempt to outline the roles and responsibilities of the Center relative to the Adaptive Management Work Group and the Technical Work Group.

Emphasis on the role of independent review in adaptive management is a second strength of the strategic plans. In addition to detailing the roles of independent review panels in proposal and document evaluation, the strategic plans provide for critical evaluation of Center programs, plans, and performance. Although some weaknesses in these independent review plans are discussed below, the Center has responded to the spirit of previous National Research Council reviews (NRC, 1987; NRC, 1996a).

A third strength is the Center's efforts to consult with stakeholders. Although these efforts have not always been entirely successful, the Center has sought stakeholder input on science plans to a greater degree and in more consistent ways than commonly occurs in such organizations, reflecting an appreciation of the aims and methods of adaptive management. Based upon the committee's observations and conversations with stakeholders, these efforts to formalize input seem to have led to greater stakeholder satisfaction than the process associated with the Glen Canyon Environmental Studies, although a systematic evaluation has not been conducted. The Center has devoted considerable energy and resources to working with stakeholders, providing scientific information, and experimenting with ecosystem research (e.g., conceptual modeling) related to adaptive management.

The introductory chapters on adaptive management in the 1998 Strategic Plan are being reworked in new documents for the overall Adaptive Management Program. If adaptive management is more fully and effectively elaborated, it will be due in part to the Center's initial efforts. Because the first Strategic Plan was initiated before the stakeholder groups were formed, and was revised as organizational roles were reexamined, it seems unlikely that the Center could have successfully articulated the full scope and nature of adaptive management. To produce its Strategic Plan, however, the Center had to try to present its science plans within the evolving context of the Adaptive Management Program. Because the strategic plans do not adequately develop some aspects of adaptive management, we comment upon some weaknesses in the adaptive management chapters and suggest alternative ways to address them.

WEAKNESSES OF THE STRATEGIC PLANS

The adaptive management chapters of the strategic plans suffer from the following weaknesses: (1) lack of clarity of the Center's roles within the Adaptive Management Program, (2) inadequate discussion of competing goals and "visions," (3) lack of clearly-defined linkages between adaptive management, ecosystem management, and social learning, (4) disparate management objectives and information needs, (5) inadequate definition of the core adaptive management experiment, (6) insufficient contingency planning, (7) insufficient decision analysis, and (8) uneven progress toward independent program review.

The Center's Roles Within the Adaptive Management Program

The division of roles and responsibilities among organizations in the Program is unclear. The Center has assumed or been charged with some administrative roles beyond those defined in the Glen Canyon Dam Environmental Impact Statement and charter documents, which may reduce its ability to perform its primary scientific tasks. At this writing, the Technical Work Group was seeking to clarify those roles.

The Center's roles were envisioned as designing and conducting research and monitoring activities to meet the needs of the Adaptive Management Work Group and the tenets of ecosystem science. The Adaptive Management Work Group is to develop and make recommendations to the Secretary of the Interior for overall management decisions and to the Center regarding management objectives related to monitoring the effects of alternative dam operations. These responsibilities were described in the original Center operating protocols as "consistent and effective cooperative efforts ongoing in the areas of policy, administrative and science protocols, definition of research needs, and dissemination of research information and technology" and as a "close functional relationship between resource stakeholders and managers and the Center's science group" (Center, 1996).

In terms of time and resources expended, the Center appears to have been highly responsive to stakeholder requests from the Adaptive Management Work Group and Technical Work Group. The Center has devoted considerable efforts to stakeholder meetings, information requests, and consultation—at the likely expense of data synthesis, integration of research programs, and implementation of monitoring programs. Center scientists have also spent considerable effort working with the Technical and Adaptive Management Work Groups in developing and revising protocols, plans, and budgets. Although these efforts may have been justified in the start-up years from 1997 to 1998, their persistence may be cause for concern. Early estimates that bimonthly Technical Work Group meetings would be needed between the biannual Adaptive Management Work Group meetings proved unrealistic, as the Technical Work Group began to meet monthly (with some ad hoc groups meeting more frequently). While necessary and valuable for coordinating management–science relationships, this time-consuming interaction between Center staff and the Technical Work Group may have delayed implementation of monitoring programs.

In addition to research and monitoring, the Center has provided organizational support and substantive assistance for the activities of both work groups. This is contrary to a model in which these stakeholder groups articulate management objectives, information needs, and a vision for the Canyon ecosystem, while the Center implements monitoring and research programs to address those management objectives, information needs, and vision. The research program has begun, but the monitoring program has experienced delays that may be partly attributable to these broader responsibilities.

Although it has helped define the overall Adaptive Management Program, the Center may become a subservient junior partner in the Program. The organizational diagram from the Glen Canyon Dam Environmental Impact Statement is triangular, suggesting an even, collaborative relationship and rough parity between the Center and the Technical Work Group (see Figure 1.2). Instead, the Technical Work Group has emerged as the implementation arm of the Adaptive Management Work Group and seems to exert a de facto review and approval authority over Center documents and budgets (authority that may have been originally envisioned for Adaptive Management Work Group). The committee is concerned that this trend may lead to micro-management and a hierarchical structure, rather than to the balanced, collaborative relationship described in the Center's original operating protocols and the Glen Canyon Dam Environmental Impact Statement. We hypothesize that a balanced, collaborative organizational structure is more conducive to the iterative and experimental aims of adaptive management than the current trend, and recommend that any changes in organizational roles be treated as experiments. Rather than exerting excessive oversight of the Center's plans and activities, stakeholders should guide the Center's scientific programs through clear management objectives and information needs.

A critical strategic issue is to ensure that a larger proportion of the Center's resources goes to monitoring and research, which depends partly upon the forthcoming Guidance Document. In articulating the Center's roles and responsibilities, that document will hopefully strike a balance between scientific work in the Grand Canyon, responsiveness to stakeholder information needs, and broader communication of scientific results (e.g., in public fora and research publications). It is important to recognize that the fiscal and human resources needed to manage a newly formed and evolving institution (such as the Adaptive Management Program) are probably greater than those required to manage a decades-old, established program. Recognition of these evolving needs would include realistic estimates of monitoring requirements—and of associated time and effort—to implement the necessary monitoring infrastructure.

Finally, an advocate is needed for the adaptive management experiments themselves, particularly regarding their scientific coherence and the long-term integrity of the Grand Canyon ecosystem. There is currently no voice among the stakeholders that represents the interests of these scientific experiments. This role might be explicitly assigned to the Center or to a senior scientist, who could help articulate and interpret scientific aspects of adaptive management within an ecosystem context.

The Concept of "Vision"

There has been limited progress to date in developing a "vision" of the desired future conditions in the Grand Canyon ecosystem. Strategic plans refer to pre-dam conditions, but pre-dam baselines are not well defined, nor is it likely that they represent desired or attainable objectives (Schmidt et al., 1998). The use of baselines may not be desirable because they do not define "endpoints," i.e., realistic and desired outcomes (cf. Dewey [1958] on "ends-in-view"). Although some scientists have begun to define promising "normative" or "naturalistic" alternatives, these have proven difficult to elaborate or implement (cf. Independent Scientific Group, 1996; Schmidt et al., 1998). Others suggest that such goals may not be necessary or as important as relative improvements in ecosystem conditions and services (cf. Brunner and Clark, 1997; Rogers, 1998).

Previous National Research Council committees also noted the lack of clear, coherent goals for Grand Canyon ecosystem management. During Glen Canyon Environmental Studies Phase I, the committee found that "The goals and objectives presented in the GCES were articulated vaguely, they were inconsistent across individual studies, and they often confused science and policy. They seemed to be more strongly related to the missions of the participating agencies than to understanding how the controlled hydrologic regime influenced downstream resources" (NRC, 1987, p. 8). The 1996 National Research Council committee noted that "The research conducted by a myriad of cooperators under the GCES umbrella would have been more effective if all the parties involved had devised a system to focus on resources on the stated purpose of the GCES... The inability of the "cooperators" to devise a system for focusing resources on the stated purpose of GCES was a remarkable and consistent feature of the program, and resulted in great expansion of expenditures and diffusion of focus" (NRC, 1996a, p. 32-33).

Some stakeholders stated that the common goals for the Grand Canyon ecosystem are approximated by the expected benefits of the preferred alternative in the Glen Canyon Dam Environmental Impact Statement (U.S. Bureau of Reclamation, 1995: see Appendix E of this report). Stakeholders generally seemed to also agree, however, that this set of expected benefits represented a political compromise. Although falling short of a coherent vision or an optimal mix of conditions, it constituted some initial "targets" for adaptive management. The Strategic Plan does not discuss the scientific implications or limitations of this mix of objectives, or the implications that a more coherent set of objectives might have. Several strategies are possible, and each strategy leads in a different scientific direction. We discuss two alternatives within a continuum of possibilities, one pluralistic, the other visionary.

The first alternative is the current pluralistic situation. Management objectives are organized under nine "resource areas" identified in the Glen Canyon Dam Environmental Impact Statement: water, sediment, fish, vegetation, wildlife and habitat, endangered and other special status species, cultural resources, recreation, and hydropower. The Glen Canyon Dam Environmental Impact Statement states that "Reasonable objectives, developed by the management agencies, are goals for future management of these resources and provide meaning to the terms 'protect,' 'mitigate,' and 'improve'" (U.S. Bureau of Reclamation, 1995, p. 54). It furthermore states "Attainment of objectives for all resources will require complex interagency planning and management. Some issues would remain unresolved under any alternative" (U.S. Bureau of Reclamation, 1995, p. 54). These appear to be the only attempts to date to define desired future conditions for the Grand Canyon ecosystem.

The Adaptive Management Program is thus left with a list of nine resource areas, each with its own logic, its own management objectives, and its own information needs. No formal attempt has been made to ascertain compatibility or incompatibility of these resource areas or management objectives, or how they may or may not fit into an ecosystem context. This matter is further complicated when the nine areas are translated into the Center's four resource program areas for monitoring and research (e.g., some resources such as "water" could fit under several program areas; see Table 3.1).

Although a strength of this pluralistic strategy is that it reflects an actual situation rather than a static baseline or utopian scheme, it has several limits. As a compromise among varied objectives, the alternatives and impacts in Appendix E reflect a "vision" of the Glen Canyon Dam Environmental Impact Statement team; however, they lack coherence and salience for many, if not all, stakeholder and scientific groups. Disparate objectives can limit stakeholders' and scientists' ability to weigh effects and alternatives as they evaluate results of various management actions. These problems may be partially alleviated by graphic visualization methods that help represent multiple-objective compromise alternatives. But they would still lack the common or prioritized objectives needed for rule-based simulation, goal programming, and optimization methods.

The second alternative—the visionary alternative—posits that until a coherent vision for the ecosystem is agreed upon, it will be difficult to create a program that meets the tenets of ecosystem science and adaptive management as specified in the Grand Canyon Protection Act, the Glen Canyon Dam Environmental Impact Statement, and the Record of Decision. To develop this visionary alternative, the Adaptive Management Work Group would need to translate the language in these policy documents, and the underlying values for which Grand Canyon National Park and Grand Canyon National Recreation Area were created, into a coherent set of objectives to design the Adaptive Management Program. The vision would need to merge concepts of conservation ecology and social welfare and test them in a context of substantial uncertainty and imperfect information. According to this thinking, until such a vision is articulated and pursued, there are no organizing principles for directing the complex research, monitoring, interpretation, and policy recommendations.

These two very different perspectives are evident among scientists and stakeholders in the Adaptive Management Program. They have stimulated vigorous, if not as yet very fruitful, debate. In this context, three recommendations may prove useful.

TABLE 3.1 Nine "Resource Areas" within the Grand Canyon Monitoring and Research Center's 1997 Strategic Plan

---

Physical Resources Program
    1. Water
    2. Sediment transport

Biological Resources Program
    3. Fishes
    4. Vegetation
    5. Wildlife and habitat
    6. Endangered and other special status species

Cultural Resources Program
    7. Cultural resources

Socioeconomic Resources Program
    8. Recreation
    9. Hydropower

The first recommendation is to start with the pluralistic situation and recognize it as the current context for experimentation with the operating alternative described in the Record of Decision. No matter how diverse the objectives, a clear statement of the current situation is needed for effective planning, implementation, and evaluation of the adaptive management experiment.

The second recommendation is that a strategy should be developed for inviting and articulating visions for the ecosystem as part of Adaptive Management Work Group and other meetings. Through time, some of these views may gain support and clarify common interests in the Grand Canyon ecosystem.

The third recommendation is that it would be wise to examine adaptive management experiments where "normative" approaches are being tried, as in the Columbia River Basin (Independent Scientific Group, 1996). It is important to recognize that issues of vision have important implications for adaptive management, including the clarification of management objectives and information needs, the conduct of science programs to address them, and ultimately trade-offs among competing objectives. The committee is encouraged by stakeholders' pursuit of these recommendations in a May 1999 river trip in the Grand Canyon.

Adaptive Management, Ecosystem Management, and Social Learning

Just as "visions" for the Grand Canyon ecosystem are at a formative stage, the Adaptive Management Program overall is at an early stage of development. It has been appropriate for the strategic plans to initially focus on establishing the Program and its protocols. It would also be useful for the plans to anticipate the more complex challenges of a mature Adaptive Management Program, especially its links with ecosystem management and social learning. Adaptive management involves a process of experimenting with management actions in the face of uncertain ecosystem outcomes. The uncertainties stem from the complexities of large-scale ecosystems, wherein it is impossible to understand all cause-and-effect interrelationships. It also involves strategic treatment of uncertainties and ecosystem management alternatives.

The strategic plans and related documents do not explicitly discuss relationships between adaptive management and ecosystem management. Instead, adaptive management currently refers first, to experiments with dam operations and their downstream resource effects; second, to the use of ecosystem science to monitor and explain those effects; and third, to adjustments in dam operations to improve the mix of effects. As the use of ecosystem science develops in the Adaptive Management Program and as a vision for downstream resources in the Grand Canyon ecosystem becomes clearer, adaptive management may evolve into a program of ecosystem management.

Ecosystem management has been defined in various ways. One reasonable working definition was provided by Moote et al. (1994): "Ecosystem management is a management philosophy which focuses on desired status, rather than system outputs, and which recognizes the need to protect or restore critical ecological components, functions, and structures in order to sustain resources in perpetuity." Machlis et al. (1997) listed five principles central to ecosystem management: (1) socially defined goals and management objectives, (2) integrated holistic science, (3) broad spatial and temporal scales, (4) adaptable institutions, and (5) collaborative decision making. They recommend a "human ecosystem" model that melds ecological and social sciences as an organizing framework. These concepts of ecosystem management have much in common with adaptive management, as developed by Holling (1978), Walters (1986), and others (Halbert 1993; Ludwig et al., 1993).

The role of "social learning" in adaptive management is implicit in the Strategic Plan, but it is not treated experimentally. Social learning occurs as stakeholders and scientists gain a clearer understanding of how the ecosystem works, how it responds to management alternatives, and how society interprets and values those responses and, on the basis of that new knowledge, makes conscious trade-offs and adjustments (Parson and Clark, 1995). The Strategic Plan recognizes processes of learning and adjustment, but it would benefit from a more scientific treatment of them. For example, the Center's conceptual modeling efforts have sought to help scientists and stakeholders learn about ecosystem dynamics, resource linkages, plausible scenarios, and surprising outcomes. Participants in conceptual modeling workshops sponsored by the Center report that the workshops were successful, but what did they learn and how did they learn it? These questions call for the same kind of hypothesis-driven analysis as the ecosystem experiments themselves. As stakeholders receive monitoring and research results, what do they learn and how do they learn it from those results? What difference does it make for their management objectives, information needs, ecosystem visions, resource valuation and, ultimately, their recommendations to the Secretary? To address some of the institutional challenges attributed to adaptive management (Walters, 1997, 1998), these basic questions of social learning should be *formally* incorporated as part of the experiment.

Disparate Management Objectives and Information Needs

As defined in the 1998 Strategic Plan, management objectives are to "define measurable standards of desired future resource conditions which will serve as objectives to be achieved by all stakeholders within the Adaptive Management Program." Information needs are to "define the specific scientific understanding required to obtain specified management objectives" (Center, 1998, p. 1, Appendix A).

The information needs are large in number (176 were in the 1998 Strategic Plan), diverse, repetitive, and sometimes conflicting and conceptually uneven. There is little evidence of cross-program linkages among specific management objectives and information needs. The main exception is "Ecosystem assessment MO 1: Develop a conceptual model of the Colorado River ecosystem." The Cultural Resources Program has close linkages with physical resources investigations. There is a clear need for a matrix or other approach that relates management objectives to one another and demonstrates their contribution to attaining some vision for and level of understanding of ecosystem conditions. Ideally, the management objectives should collectively represent stakeholders' vision for the Grand Canyon. Instead, the management objectives and information needs reflect the lack of a coherent vision.

Information needs vary across programs. The Cultural Resources Program's information needs are generally clear and manageable in number. Some of the Physical Resources Program's information needs are imprecise and difficult to understand. Many of the Biological Resources Program's information needs focus on avoiding jeopardy for threatened and endangered species, rather than on developing understanding of the ecosystem. There is also overlap between information needs, both within and across resource categories. A lack of clarity in individual information needs works against the stated purpose of clearly communicating research and monitoring aims. At a more fundamental level, the lack of a clear set of management objectives and information needs makes it difficult to design and test adaptive management experiments.

The management objectives were derived from the Glen Canyon Dam Environmental Impact Statement and the resource areas defined therein. Although a Technical Work Group subgroup of 14 members worked to revise the information needs in April 1998, the resulting list had many similarities to the original set developed in 1996 from the Glen Canyon Dam Environmental Impact Statement. The subgroup attempted to prioritize information needs to provide guidance for the timing of research and monitoring programs. But if information needs overlap or are poorly worded, ballots of this sort would be flawed, providing a questionable basis for prioritizing research and monitoring projects.

Physical Resources Program

In the Physical Resources Program, management objectives were defined by a stakeholder working group in 1996. Essentially, the same management objectives reappear in both strategic plans (see Table 3.2).

These management objectives state that the number and size of sandbars and backwaters should be maintained, that sediment should be redistributed from the channel bottom to the channel margins, and that there be enough sediment in the channel to support this process. Although

TABLE 3.2 – Physical Resources Program Objectives

1.   1998 **Management Objective 2** Maintain at 1990–91 levels:
      (a) the number and size of sandbars between the 8,000
      and 45,000 cfs stages
      (b) the number and size of backwaters at the 8,000 cfs
      stage

2.   1998 **Management Objective 3** In as many years as reservoir
      and downstream conditions allow, increase:
      (a) the size of sandbars above the 20,000 cfs stage and
      (b) the size of backwaters at the 8,000 cfs stage to levels
      observed following the 1996 BHBF (beach/habitat-
      building flows)

3.   Maintain "system dynamics and disturbance" by redistributing
      sand stored in the main channel and eddies to areas inundated by
      flows:
      (a) (1998 **MO1**) between 20,000 and 30,000 cfs in years
      when Lake Powell water storage is low and
      (b) (1998 **MO4**) up to 45,000 cfs in years when Lake
      Powell water storage is high and downstream resources
      warrant.

      Monitor these targets by measuring:
      (a) the area of bare sediment deposits and
      (b) the number and size of representative sandbars.

4.   1998 **Management Objective 1** Maintain a long-term balance of
      river-stored sand to "support" high flows (annual habitat
      maintenance flow, beach/habitat-building flows, unscheduled
      flood flows).

SOURCE: Center (1998).

basically correct, the management objectives are repetitive and unnecessarily vague. The first two state basically the same thing: that the number and size of sandbars and backwater channels be maintained. These could be combined and stated more concisely and clearly. The third management objective listed in Table 3.2 requires that "system dynamics and disturbance" be maintained, while the fourth requires maintenance of "a long-term balance of river-stored sand to support high flows." These objectives are unnecessarily vague and could be stated in simpler language. Some of the information needs developed for the sediment resource program are similarly repetitive, imprecise, and difficult to understand.

Biological Resources Program

The management objectives and information needs for the Biological Resources Program are also unwieldy and repetitive. There are 16 management objectives listed for biological resources. Within each of these, three to eight information needs are identified. Although identifying many needs may be a helpful starting point, they must be consolidated and reduced in number before a scientific plan for addressing them can be established.

Management objective 2 (Table 3.3) provides an example of how some consolidation could be accomplished. It states that "downstream of Glen Canyon Dam to the confluence of the Paria River, sufficient ecological conditions should be maintained...to produce a large, self-sustaining population of at least 100,000 Age II+ rainbow trout." Information need 2.1, "Determine ecosystem requirements, population character and structure to maintain naturally reproducing populations of Age II plus fish at 100,000 population levels in Glen Canyon," is an overarching statement that captures the essence of what is desired. The next five information needs (2.2–2.6) express details of information to be gathered to fulfill information need 2.1. Information need 2.5 seems redundant with 2.1. Information need 2.7 is redundant with 1.3. To make the information needs conceptually parallel, they either need to be collapsed into a single, integrated information need similar to 2.1, or 2.1 should be eliminated and the detail in information needs 2.2–2.6 maintained.

Sociocultural Resources Program

The management objectives and information needs for the Cultural Resources Program are clear and coherent. By contrast, management objectives and information needs for the Socioeconomic Resources Program contain major omissions, elaborated in Chapter 4 and

TABLE 3.3 – Biological Resources Program Objectives

1.   **1998 Management Objective 1** Maintain and enhance the aquatic
       food base in the Colorado River ecosystem to support
       desired populations of native and non-native fish.
       Information Need 1.3
              Determine the aquatic food base species composition,
              population structure, density, and distribution required to
              maintain desired populations of native and non-native fish
              in the Colorado River ecosystem.

2.   **1998 Management Objective 2** In the Colorado River
       downstream of Glen Canyon Dam to the confluence of
       the Paria River, sufficient ecological conditions (such as
       habitat, food base, and temperature) should be maintained,
       which in conjunction with management by Arizona Game
       and Fish will produce a healthy self-sustaining population
       of at least 100,000 Age II+ rainbow trout that achieve 18
       inches in length by Age III with a mean annual relative
       weight of at least 0.90.
       Information Need 2.1
              Determine ecosystem requirements, population character
              and structure to maintain naturally reproducing
              populations of Age II plus fish at 100,000 population
              levels in Grand Canyon.
       Information Need 2.2
              Determine trends in rainbow trout population size,
              character and structure in Glen Canyon.
       Information Need 2.3
              Evaluate harvested and field sampled rainbow trout to
              determine the contribution of naturally reproduced fish to
              the population in Glen Canyon.
       Information Need 2.4
              Determine the availability and quality of spawning
              substrates in the Glen Canyon reach necessary to sustain
              the rainbow trout fishery.

continues

TABLE 3.3 Continued

Information Need 2.5
> Determine the growth and condition of rainbow trout in Glen Canyon.

Information Need 2.6
> Define criteria (e.g., temperatures, flow regimes, contaminants, metals, nutrients) for sustaining a healthy rainbow trout population in Glen Canyon.

Information Need 2.7
> Determine the trophic relationship between trout and the aquatic food base including the size of the aquatic food base required to sustain the desired trout population in Glen Canyon.

SOURCE: Center (1998).

Appendix F, making it unclear how this latter program is to meet the goals of adaptive management.

Information Technology Program

Management objectives and information needs in the Information Technology Program also appear to be well-organized and internally consistent.

A simpler set of management objectives within a consistent ecosystem vision is needed for the Adaptive Management Program. A mechanism is needed for effectively revising and consolidating the management objectives and information needs to make a clear statement of desired future conditions and to provide a basis for formulating adaptive management experiments. It is thus recommended that the Center, along with a newly designated senior scientist, work with the Technical Work Group to reformulate management objectives and information needs and place them within an internally consistent ecosystem context. The revised management objectives would be based on existing management objectives and would be submitted to the Adaptive Management Work Group for consideration. The intent of this recommendation is not to move authority for defining management objectives to the Center. Rather, it is to assign the Center the task of translating stakeholder objectives into scientific needs that are clear and internally consistent and that fully incorporate an ecosystem view of the Grand Canyon. The ecosystem view would likely identify information needs that cut across and help integrate management objectives.

Defining the Current Experiment

Within the strategic plans and Program documents, this committee found no clear statement of the current adaptive management experiment. As mentioned, this experiment is based on the Modified Low Fluctuating Flow regime and beach/habitat-building flows specified in the Record of Decision. Even if management objectives and information needs are less consistent and less clear than desirable, the clearest possible statement of the current experiment is necessary. Without it, it will be difficult to develop informed opinions about outcomes and tradeoffs, and difficult to develop effective or appropriate follow-up studies.

The strategic plans do not elaborate the process for using an experimental approach as a part of the management process. Although there are discussions of this issue in guidance documents and plans, the principles of science-based management and ecosystem integration are not consistently used. Management actions (e.g., flow rate, controlled releases, and temperature of dam releases) are experiments that should have clearly defined hypotheses regarding expected outcomes across resource areas and the ecosystem. Despite this stated approach, there is no consistent presentation of hypothesis-testing in or across resource program activities. The Center's recent dialogue with the Technical Work Group on experimental design of alternative beach/habitat-building flows provides a model that has broader application (Argonne National Laboratory, 1999; Melis et al., 1998; Ralston et al., 1998). This hypothesis-testing approach is an essential component of adaptive management.

Although the current management experiment is mandated in the Record of Decision and described in some detail in the Glen Canyon Dam Environmental Impact Statement, an explicit discussion of this experiment does not appear in the Strategic Plan. A set of multiple hypotheses about anticipated outcomes of the current Modified Low Fluctuating Flow experiment should be constructed for each of the nine resource areas. As these hypotheses are tested, the Strategic Plan should indicate how the monitoring and research programs will determine if predicted outcomes occur. It should also provide for periodic discussion of alternative management experiments.

Contingency Planning

The 1997 Strategic Plan included a section on contingency planning (e.g., for unanticipated hydrologic events that could trigger beach/ habitat-building flows). As noted earlier, contingency planning can be closely related to theories of "surprise" in adaptive management. In addition to hydrologic events, surprises may include climate anomalies (e.g., El Niño), policy changes (e.g., National Park Service wilderness designation for Grand Canyon National Park), and social changes (e.g., in environmental or economic values). While some of these contingencies may be anticipated and planned for as possibilities, others may be entirely unexpected. A monitoring and research plan for adaptive management should include the latter as well as the former.

To its credit, the Center already includes contingency planning for beach/habitat-building flows in its research contracts. However, the 1998 draft Strategic Plan mentioned contingency planning in its executive summary but not in the body of the Plan or budget. The next Strategic Plan should explicitly address contingency planning and, in the spirit of adaptive management, strive to anticipate a broad range of ecosystem and societal "surprises" that could substantially affect scientific monitoring and research.

Decision Analysis

Adaptive management ultimately involves difficult trade-offs among competing objectives. The Strategic Plan concentrates on quantifying physical, biological, cultural, and conventional market economic consequences of dam operations (using incommensurate units). The Strategic Plan sidesteps the final, equally essential, step of management, the articulation of scientific criteria to guide choices among competing objectives that "protect, mitigate adverse impacts to, and improve the values," identified in the Grand Canyon Protection Act. Although those criteria and choices rest with stakeholder groups, the Center should develop scientific decision support systems to support those efforts.

The conceptual modeling effort supported by the Center and the Adaptive Management Program is an important first step in addressing the complexity of issues and potential decision scenarios related to the impacts of dam operations. To ensure that the Program is a working example of complex resource management and policy, additional decision analysis capabilities should be developed. Given the amount and sophistication of data analysis required, these are generally computer-based models referred to as decision support systems.

Reitsma (1996) described a promising approach to decision support system applications to resource management, using the Annual Operating Plan for the Colorado River as a case study. That study indicated that decision support systems should be formally implemented at several levels of decision-making, which have several parallels with the Adaptive Management Program. For example, the decision support system might include three major components:

1. State information. State information includes data representing an ecosystem's state at any time. It would include data on Glen Canyon Dam operations, power production, stream hydrology and geomorphology, temperature and oxygen concentration, aquatic primary production and detritus, benthic insect production, riparian vegetation, and animal populations. Reitsma (1996) notes that "State representation by means of databases forms the heart of modern decision support systems" because it files, retrieves, manipulates, and displays information with modern relational database management and geographic information systems.

2. Process information. Process information includes principles that represent the dynamics of various resources. Work currently underway on the Grand Canyon ecosystem model, when complete, will provide stakeholders, scientists, and the public with opportunities to qualitatively evaluate impacts of proposed actions without actually modifying the system. Instead, initial conditions, boundary conditions, parameters, and the configuration of physical submodels can be modified to assess proposed changes.

3. Evaluation tools. Evaluation tools permit quantitative and qualitative analysis and visualization of alternative actions.

The components of a decision support system are or soon will be available at the Center. Only the "glue" that might hold them together is missing. If the Information Technology Program fills its assigned role as data manager, state information will be available soon. If the Socioeconomic Resources Program fulfills its roles, it could develop a capability to quantify the efficiency of proposed dam-operation alternatives in terms of power revenue, white-water rafting, campsite availability, trout availability, and nonmarket values.

These actions could enable the Center to concentrate on scientific aspects of policy experiments and develop expertise in objective measurement of social values of the Grand Canyon's nonmarket environmental goods for different stakeholder groups. Each organization in the Adaptive Management Program could benefit from decision support tools that may lead to formal decision support systems for different parts of the Program. The committee recognizes the scientific difficulties and political sensitivities of these tasks, underscoring the importance of maintaining high standards of independent review.

Independent Review

Independent review played an important role in evaluating, and at times redirecting, the Glen Canyon Environmental Studies. The previous National Research Council committees ran from 1986 to 1987 and from 1991 to 1996. Important changes in the Glen Canyon Environmental Studies were made in direct response to some of the committee's recommendations, including the establishment of an office of senior scientist (based in part on the recommendation of the first National Research Council review), consideration of nonuse values, analysis of power economics, and reevaluation of Lake Powell evaporation. Glen Canyon Environmental Studies underwent fundamental changes during the committee's tenure (NRC, 1996a), including evolution of a framework for administrating science and for monitoring and incorporating scientific information in the policy process.

This committee is presented with a very different organization, one that is more complex and situated within a more formal stakeholder organization. The Center and Adaptive Management Program are as yet not settled or fixed. The evolving nature of ecosystem science and management and the interactions between the Center and the Adaptive Management Program argue for continuation of external review.

External Review

The Glen Canyon Dam Environmental Impact Statement includes independent review panels as a component of the Adaptive Management Program and states that, "All monitoring and research programs in Glen and Grand canyons should be independently reviewed" (U.S. Bureau of Reclamation, 1995). According to the Glen Canyon Dam Environmental Impact Statement, independent review panels are to be comprised of qualified individuals not otherwise participating in monitoring and research studies and established by the Secretary of the Interior. Furthermore, these panels are to be established in consultation with the National Academy of Sciences (parent body of the National Research Council) and the Adaptive Management Work Group. Review panels would be responsible for periodically reviewing resource-specific monitoring and research and for making recommendations to the Adaptive Management Work Group and Center regarding monitoring, priorities, integration, and management (U.S. Bureau of Reclamation, 1995, pp. 37-38 and Figure II-10). Specific responsibilities of the review panels include annual review of the monitoring and research program, technical advice requested by the Center or Adaptive Management Work Group, and five-year review of monitoring and research protocols.

The letter that founded the Center stated that its annual funding was to be proposed by the Center chief after consultation with the Adaptive Management Work Group *and* an independent scientific review panel (Deputy Assistant Secretary for Water and Science, 1995). The Center's operating protocols (Center, 1996) provide guidelines for review of short- and long-term science plans, monitoring and science proposals, data, research reports and publications, and general program accomplishments. These guidelines include independent review of the long-term monitoring and research plan by the National Research Council, which is to interact with the Center, the Adaptive Management Work Group, and the Technical Work Group in providing guidance on the Strategic Plan.

Monitoring and research within the Adaptive Management Program can benefit from external review at three levels:

1. Proposals and reports. This level of review is in place and operating effectively. Proposals and reports are mail reviewed by external experts, with Center program managers coordinating the reviews. Review panels are convened to provide collective judgment on proposals. Current review procedures are carefully defined and are reasonable and effective.

2. Review of resource programs. A panel of experts within the domain of each program or technical area reviews research projects and advises on program direction. The Center's protocol evaluation program provides some but not all of this function. Within the protocol evaluation program, all resource programs will be reviewed over a three- or four-year cycle. In fiscal year 1998–1999, remote sensing and physical resources underwent protocol evaluation program reviews; biological and cultural resources will be reviewed in fiscal year 1999–2000.

The scope of the protocol evaluation program is limited to determining "the most effective and feasible methods of measuring Colorado River Ecosystem resource attributes and their long-term responses to GCD [Glen Canyon Dam] operations under the ROD [Record of Decision]" (Center, 1996). If strictly interpreted, this scope does not encompass full programmatic review. In addition to evaluating whether the best methods are used, external review panels should also be encouraged to evaluate whether the best questions are being asked. The scope of the protocol evaluation panels should be broadened to encompass unrestricted review of each program. Such review would include protocol evaluation and also broader questions of objectives and coordination.

3. Review of the Center and the Adaptive Management Program. In addition to review of individual programs, the Center and the Adaptive Management Program will benefit from review of overall monitoring and research and its effectiveness in addressing the mandates of the Grand Canyon Protection Act, the Glen Canyon Dam Environmental Impact Statement, and the Record of Decision. A multidisciplinary committee is essential for adequate consideration of coordination and balance among resource programs, their combined effectiveness in advancing understanding of the Grand Canyon ecosystem, and progress in defining and testing adaptive management experiments.

Programmatic Review

Although proposal, report, and resource program review activities are currently effective and should continue, the format and responsibilities for broad programmatic review still need resolution. The Center has proposed the creation of a Science Advisory Board, which could fulfill this broad programmatic review function. In the recent request for proposal for membership, the initial activities included review of requests for proposals, annual plans, and budget priorities. In a final discussion paper (dated

March 17–18, 1998; adopted by the Adaptive Management Work Group on July 21, 1998), review responsibilities of the Science Advisory Board included a five-year review of monitoring and research protocols and the long-term monitoring plan.

As currently proposed, the Science Advisory Board would face several constraints that may inhibit its ability to provide truly independent review. Part of the Science Advisory Board's proposed role is to provide scientific advice as needed by the Adaptive Management Work Group, the Center, or the Secretary of the Interior. The Science Advisory Board's ability to provide unbiased criticism may be compromised if it has an influence on the types of projects conducted and the methodology used to conduct them. Such a problem was noted by a previous National Research Council committee (NRC, 1996a), which played a dual role in advising on projects and critiquing them. A second obstacle to independent review relates to institutional constraints. According to the March 17–18, 1998 discussion paper, the Science Advisory Board is to be an official subcommittee of the Adaptive Management Work Group. The paper goes on to instruct the Science Advisory Board to "not review, interpret, or otherwise evaluate public policy decisions...associated with the Glen Canyon Dam Adaptive Management Program and activities of the AMWG [Adaptive Management Work Group], the TWG [Technical Work Group], or individual member agencies." These formal constraints, particularly when combined with its "in-house" advisory role, would compromise the Science Advisory Board's ability to provide thorough, rigorous, and unbiased external review. Although review of public policy decisions and legal compliance may not be the principal charge to a review panel, such explicit limitations are neither appropriate nor productive.

In the current solicitation, Science Advisory Board members are self-nominated or are nominated by stakeholders. Neither of these methods would promote the perception of independent, unbiased review. These methods of solicitation have also proven ineffective in attracting a pool of applicants with acceptable qualifications. These limitations and its proposed subcommittee status suggest that the Science Advisory Board as defined is not likely to provide the required kind of external, independent review.

The format and responsibilities for broad programmatic review must be resolved. Only one body should conduct such review. More than one review body would be inefficient and expensive and would place an unfair burden on the Center staff, who would have to respond to two review bodies and could well get caught between them. If the Science

Advisory Board is to be used for broad programmatic review, a number of changes are required to ensure credibility and independence. It should not be defined as a subcommittee of the Adaptive Management Work Group, which would make it an internal organization. Formal constraints should not be placed on the range or kind of issues that it may consider. Although the Science Advisory Board may be asked to focus on particular issues, it should also be free to comment on broader aspects of those issues. Its membership should be by invitation, with selection determined by Center professional staff with consultation from an ad hoc external scientific advisory group. Finally, the Science Advisory Board's advisory roles should be clarified to minimize potential conflict between advice and criticism.

4

Ecosystem Monitoring and Science

The Center has adopted an ecosystem approach to understanding the effects of Glen Canyon Dam operations on the Grand Canyon. This chapter thus begins with comments on ecosystem studies and monitoring, then reviews the Center's physical, biological, cultural, socioeconomic, and information technology programs. As pointed out in previous National Research Council reviews (1987, 1996a), an ecosystem approach seeks an understanding of interrelationships among important physical, chemical, biological, and social processes. Here we evaluate the Center's progress toward planning and implementing an integrated and comprehensive ecosystem-level monitoring and research program. In particular, two key components are evaluated: development of a conceptual model of the Grand Canyon ecosystem and the long-term monitoring program.

Much of the Center's efforts in these areas build upon earlier programs and data gathered by the Glen Canyon Environmental Studies. The Center's use of GCES data and methods is uneven, ranging from good use of past physical sciences and cultural studies to little use of past work in studies on socioeconomic values of resources. The Center has assembled a large amount of information from Glen Canyon Environmental Studies, however, including synthesis projects to determine the limits of those data and methods. This chapter discuss instances in which data from the Glen Canyon Environmental Studies have proven useful for the Center's resource programs.

CONCEPTUAL MODELING

The conceptual model was specified in the 1997 Strategic Plan and the 1998 Strategic Plan. Development and analysis of the conceptual model is the most tangible evidence that the Center is advancing concepts of ecosystem science and management toward a perspective of how alternative dam operations affect downstream resources that is integrated across physical, biological, and social science disciplines.

While no single model will capture all processes important to Grand Canyon resources, the Center's efforts in conceptual modeling have helped draw together previously disparate and independent data sets. The Center has built upon Glen Canyon Environmental Studies' conceptual models that were not computerized, and has provided a forum for discussion and interaction among stakeholders and scientists of diverse disciplines. The model and a 1999 Colorado River Ecosystem Science Grand Canyon Symposium are helping integrate the scientific thinking of Center staff and other scientists working in the Grand Canyon. That the Center was able to implement a modeling exercise with leading experts in the field (Korman and Walters, 1998) is encouraging evidence that it is capable of overseeing an excellent ecosystem-level science and monitoring program.

Conceptual modeling is proceeding on a reasonable schedule, with the initial contract likely to be completed in fiscal year 1999. Although the original Strategic Plan indicated continuing efforts to refine the model, based on future monitoring and research, there is no firm evidence in the 1998 Strategic Plan of continuing model development. It is anticipated that the most useful outcomes from the current modeling effort will be the identification of key ecosystem uncertainties and stimulation of discussion and action regarding data shortcomings. For example, one weakness identified in the exercise is a lack of long-term comparable data on trends in native and nonnative fish populations. Other preliminary results suggest that interim flows may have been beneficial to some fishes due to increased primary production in the Lee's Ferry reach of the Colorado River. They also indicate that the predam ecosystem may not have supported a great abundance of native fishes. It seems clear that the model already has been useful in framing important ecosystem-level questions.

The conceptual model project is not, however, designed to address all questions of interest to the Center. For scaling reasons, some processes at fine spatial or temporal scales are not included, such as modeling the dynamics of individual sandbars critical to understanding processes leading to their erosion and development. For this and other excluded processes, the development of separate, more focused models will be required. Some socioeconomic data have been included, but not as systematically as ecosystem data, and cultural resources had not been incorporated at the time of this review. The Center rightly emphasizes that the conceptual model should not be viewed as a predictive tool. Its primary value is obtained through its construction, which can help guide further studies, rather than its specific predictions. For similar reasons, the Center has rightly cautioned stakeholders that the conceptual model is not a decision-making tool. However, development of a new decision support system could certainly build upon lessons learned in conceptual modeling.

The model's development should be viewed as an early and significant success, and the Center should be encouraged to use the exercise and its methodology as a vehicle for integrating future programs of science, monitoring, and adaptive management. Improvements in the conceptual model of the Colorado River ecosystem represent an important step forward, as synthesis and integration are areas where Center programs lag behind the goals expressed in the original Strategic Plan.

DESIGN AND IMPLEMENTATION OF A LONG-TERM MONITORING PLAN

The past fifteen years of research in the Grand Canyon have left a mixed legacy. On one hand, there has been progress in understanding sediment movement, the effects of water-level fluctuations on some aspects of benthic community dynamics, short-term responses to an experimental controlled flood (AGU, 1999), and other issues. On the other hand, there is still inadequate understanding of how long-term physical and biological dynamics are affected by dam operations. There are relatively few internally consistent, long-term data sets that span these fifteen years. Such data sets are needed to develop a comprehensive understanding of how variations in dam operations have affected Grand Canyon resources (for recent syntheses see Grams and Schmidt, 1999; Patten, 1998; Valdez and Carothers, 1998).

One reason there are few long-term data sets useful in quantitative assessment of ecological changes in the Grand Canyon is that a long-term monitoring strategy and plan were not developed and implemented for reasons reviewed in Chapter 1. The Center has correctly identified the need for a scientifically sound, comprehensive, long-term monitoring program as a major priority. The Strategic Plan discusses many principles on which a sound, long-term program should be based. These include analysis and synthesis of existing data, development of a conceptual ecosystem model, the need to be conservative in modifying a monitoring program once started (both in terms of items monitored and methods used), and provision of an information management system capable of safeguarding and assuring easy access to long-term data. The Center has also developed an Integrated Water Quality Program, which builds upon monitoring activities initiated in the Glen Canyon Environmental Studies period (Vernieu and Hueftle, 1999). The Integrated Water Quality Program encompasses quarterly reservoir surveys, monthly forebay surveys, and selected downstream monitoring. It uses the Lake Powell split criteria to indicate which monitoring activities—and associated management objectives and information—fall into "white," "gray," and "black" categories. It also specifies sampling locations, frequencies, and analysis.

The Center is clearly aware of the many issues that must be considered in designing a successful long-term monitoring program. The committee is concerned, however, that in contrast to the excellent materials in the Strategic Plan regarding principles of monitoring, there are few details about the emerging monitoring plan itself, or about application of these concepts to the Grand Canyon ecosystem. The Strategic Plan falls short in its lack of discussion of the major next steps toward implementing long-term monitoring. For example, with the exception of the Integrated Water Quality Program, there are no tabulations of existing long-term data sets, no tentative lists of variables that might be considered for measurement, and little mention of where within the ecosystem it may be best to make measurements. The Strategic Plan calls for protocol evaluation programs to critically evaluate sampling protocols proposed by each resource group, but it is unclear if there is a mechanism to ensure integration across resource groups.

The Center should place a high priority on developing a detailed, long-term, integrated monitoring plan. The lack of a plan will hamper the rest of its functions, including development of requests for proposals. The monitoring plan must be designed to provide data necessary to evaluate long-term responses to current and future adaptive management. While the flows prescribed in the Record of Decision are now the main adaptive management experiment, the range of alternatives considered will likely broaden over time. There are, for example, at least two additional adaptive management experiments currently under consideration: short-term beach/habitat-building flows and installation of a temperature control device at Glen Canyon Dam. In the long term, it is likely that other management options not currently envisioned will become available. Perhaps the only way to ensure that a long-term monitoring program will be relevant to evaluating the broad suite of experiments that may be conducted is to adopt a long-term ecosystem-level perspective. The following suggestions are offered in support of the Center's efforts:

• A long-term view of the monitoring program should be adopted. Long-term monitoring often yields few benefits in the first several years. A program designed to detect long-term (five- or ten-year scales to a multidecadal scale) changes should not be expected to yield significant results in the first few years. A lack of short-term results must not be allowed to impede development and implementation of a long-term program. Some of the more effective long-term data sets consist of rela-tively simple variables whose values accrued because of long-term sampling. An excellent example of a simple, yet powerful, long-term data set is the Secchi disk record collected since 1967 at Lake Tahoe, California (Jassby et al., 1999), which has documented slow but definite reduction in water clarity related to biologic responses to increased nutrient inputs from the watershed.

• Because ecological processes operate over various temporal and spatial scales, a long-term monitoring program should be effective at several different scales. Focusing evaluation of processes at a single spatial or temporal scale may result in an overly narrow view of Grand Canyon dynamics. The Center should consider a hierarchical design, consisting of a few local sites monitored frequently in detail, several index sites that receive less detailed monitoring at longer intervals, and broader reaches that might be monitored least intensively, perhaps using airborne (or other) remote sensing at annual or longer frequencies.

• The core variables forming the basis of the monitoring program should be explicitly identified. Core data sets should consist of simple, basic data whose value will accrue over time. Core data sets should be selected using an ecosystem-level, multispecies perspective, ensuring salience of variables over the long-term. Even at this early stage, there should by now be an identified list of candidate variables and measurement locations, frequencies, and methods. It is troubling that a preliminary listing is not in the 1998 Strategic Plan.

• Once established, the monitoring program must be protected from fluctuating budgets and changing short-term interests. A monitoring program's value is in its long-term nature. Mechanisms must therefore be developed to buffer it from short-term fluctuations in the Center's budget. A consensus should be developed among scientists and stakeholders that the monitoring program receives first priority in lean budget years.

• Short-term research projects must be closely linked with the monitoring program. These short-term research projects should be identified by scientists and can be overseen by the stakeholder groups. Prospective short-term research projects should be partly evaluated in the peer-review process by their likelihood of providing a better understanding of relationships among or within the core monitoring data. Examining how short-term projects will enhance understanding of linkages between and among long-term data sets can provide an important way to focus research toward the needs of adaptive ecosystem management.

• Physical, biological, cultural, and socioeconomic measures should be co-located in space and time wherever and whenever practical. Co-location of monitoring variables, sites, and times among programs is an excellent way to assure integration across resource groups (such as monitoring of the controlled flood event in 1996). It may be increasingly important as the Grand Canyon National Park implements wilderness and other research permit regulations. While co-location is not always possible, there should be strong reasons before making the decision not to measure variables across resource groups at the same place or time. Optimizing co-location of sampling sites requires that the monitoring program for each resource group be developed in parallel with mechanisms for meaningful interactions among groups. The Physical Resources Program has made significant progress toward a long-term monitoring plan and has already convened a meeting of its protocol evaluation program team. The committee is concerned that because other resource groups are behind the physical group in planning and implementation, it will become increasingly difficult to develop integration across groups.

The Center should also ensure that a search for a perfect monitoring plan does not become the greatest impediment to implementing an effective long-term monitoring program. It is understood that no long-term monitoring program will be able to measure all the important variables with the frequency and spatial coverage that might ultimately be desired. Every program is thus open to valid criticism that it does not measure one or more important variables. The Center must avoid making the long-term monitoring program so ambitious and complex that it is too unwieldy to implement.

The Center should consider designing the monitoring program in stages. With each resource program using the conceptual model and with clarified information needs as a framework, the Center might wish to draft a comprehensive list of candidate variables. It could then assign variables to one of several priority lists and begin a process of determining acceptable frequencies and measurement locations. For some variables this will be procedurally straightforward; for others it will become clear that methods are inadequate or benefits of measurement are unclear, and a decision to include it will be held in abeyance until more information becomes available. Through iteration at both the individual resource group and across resource group levels, a draft monitoring plan will emerge. Involving protocol evaluation program teams for each resource group should be encouraged early in this process.

THE CENTER'S RESOURCE PROGRAM AREAS

Physical Resources Program

Management options for addressing downstream impacts of the Glen Canyon Dam are defined primarily in terms of physical controls: flow rates and temperatures of water released at the dam. Adaptive management experiments intended to improve ecosystem resources are linked to dam operations through processes of water flow, sedimentation, and erosion. A description of the physical responses of the Grand Canyon to past and future dam operations provides the framework needed to formulate adaptive management experiments and test hypotheses regarding ecosystem responses to dam operations.

A primary focus of the Physical Resources Program is sand within the Grand Canyon and its sources, sinks, and rates of transport. Sand deposits form camping beaches, provide terrestrial and aquatic habitat, and preserve cultural artifacts. Research and monitoring are focused on understanding how to maintain adequate volumes and appropriate morphology of these deposits in order to preserve associated ecological, recreational, and cultural resources. A sand budget quantifying inputs, storage, transfers, and output provides the conceptual framework for most sediment research in the Grand Canyon. Individual projects focus on inputs from gauged and ungauged tributaries, transport rates within the mainstem, and changes in storage within the channel and along its margin. A budget focuses attention on the large post-dam decreases in sand supply and the need to carefully manage the available sediment.

A second focus of the Physical Resources Program is on coarser sediments (cobbles, boulders) that form debris fans at tributary canyons, creating rapids and anchoring most of the larger sandbars in the mainstem. The post-dam flow regime has reduced the river's ability to rework these debris fans. A better understanding is needed of the ability of available floods to rework these deposits and maintain navigability of the rapids.

Synthesis of Previous Knowledge

Evaluation of past data and research is an active part of the Physical Resources Program, and the committee noted that this program was actively and carefully reviewing and building on past research. The 1998 Strategic Plan includes two research efforts that reanalyze existing data sets for the purpose of developing a consistent historical record of sand storage and transport. One is a compilation of past observations of sandbar volumes. Work conducted at various times by different organizations using a variety of methods has produced historical data on sand bar changes that are difficult to compare and interpret (Grams and Schmidt, 1999). The ability to predict future changes in sand bars will clearly benefit from a better understanding of their history. A second project is reanalyzing historical records of sediment transport in the Grand Canyon and its immediate tributaries. This work has improved understanding of sand transport before and after the dam (Topping et al., 1999) and has contributed to a fundamental revision of the sand budget and a reevaluation of the frequency and timing of floods that would best conserve sand resources.

The sand budget paradigm has provided a consistent organizing concept for sediment research over two decades (e.g., Howard and Dolan, 1981; Randle et al., 1993; Schmidt, 1999; Smillie et al., 1993, cited in U.S. Bureau of Reclamation, 1995; Topping et al., 1999). Revisions in the sand budget reflect important advances toward understanding and managing sand resources in the Grand Canyon. A revision currently under investigation is the channel's ability to store tributary-derived sediment, which has important implications regarding the timing of controlled floods needed to preserve available sand. While the Glen Canyon Dam Environmental Impact Statement was being written, it was believed that tributary sand was stored in the channel in years without large dam releases, leaving it available for redistribution to bars and channel margins by occasional controlled floods. This model was based on sand budgets developed from U.S. Geological Survey gauging records and was based on the assumption that relations between sand transport and discharge were stable over time (Randle et al., 1993; Smillie et al., 1993, cited in U.S. Bureau of Reclamation, 1995). U.S. Geological Survey cross-sections of the Colorado River were used in determining sand storage in the channel, information important to planning the controlled flood of 1996. Reanalysis of sediment gauging records (Topping et al., 1999) and observations during the 1996 controlled flood (Rubin et al., 1998; Smith, 1999; Topping et al., 1999) indicated the concentration and size of sediment transported at a given discharge can vary depending on the duration of mainstem flows and their timing relative to tributary sediment inputs. The existence of previously assumed multi-year in-channel storage is now in question, raising important new questions concerning the effective timing and duration of future controlled floods.

A previous National Research Council committee recommended several areas of research and monitoring to support management of the sand resource, including developing triggering criteria and flow specifications for beach/habitat-building flows, monitoring rates of beach deposition during beach/habitat-building flows, and creating a procedure for determining sand budgets for different parts of the Grand Canyon (NRC, 1996a). Research and monitoring supporting all of these recommendations is ongoing, and much of it is incorporated in the 1998 Strategic Plan. Results of ongoing work in each of these areas are also being used to evaluate and revise management decisions. Studies of beach deposition during the 1996 beach/habitat-building flows (Andrews et al., 1999; Center, 1997a; Hazel et al., 1999; Kearsley et al., 1999; Schmidt et al., 1999b) and research on channel-eddy sand exchange (Rubin et al., 1998; Smith, 1999; Topping et al., 1999; Wiele et al., 1999) contributed directly to ongoing discussions of the most effective magnitude and duration of such management events.

Beach/habitat-building flow triggering criteria have been developed (Technical Work Group, 1997), and information produced by the Physical Resources Program is being used to evaluate the combination of beach/habitat-building flow magnitude, duration, and post-flood flow regime that will provide the best test of the effectiveness of such efforts in conserving sand. Although a sand budget is not yet complete, ongoing research is effectively focused on components that are the least understood.

Likely Effectiveness of the Strategic Plan

Progress in developing understanding of the physical behavior of the Colorado River in the Grand Canyon is evident in the revision of the Strategic Plan. The 1998 Strategic Plan focuses attention on river reaches nearest the dam, where impacts of post-dam reductions in sediment supply are largest. Reaches in Glen and Marble canyons are considered critical because they have little sand input and have shown progressive loss of stored sand in the post-dam period (Schmidt and Graf, 1990; Schmidt et al., 1995; Webb, 1996). The long-term volume of sediment that may be stored in Glen and Marble canyons, its variability in space and time, and, therefore, the viability of related biological and recreational resources, remains to be determined. The 1998 Strategic Plan identifies needs for greater understanding of sand storage potential and sediment residence time in Marble Canyon. The 1998 Strategic Plan places increased emphasis on a fine-grained sediment budget as the primary organizing principle for continued research and monitoring. A sand budget serves to focus attention on parts of the system for which understanding is weakest (e.g., storage and evacuation of sand on the channel bed), while also supplying an internally consistent accounting as a strong basis for long-term monitoring.

The 1998 Strategic Plan also emphasizes the need for a complete map of topography and sediment content of the river corridor from the channel bed up to pre-dam flood elevations. Such a map will provide the basis for accurate routing of flow and sediment through the canyon and gives a baseline for effective, long-term monitoring of sediment. The need for a synoptic channel sediment map was recognized by a review panel convened in August 1998, and the Physical Resources Program responded within a month with an end-to-end (from Lee's Ferry downstream to

Diamond Creek, located at River Mile 225), side-scan sonar survey of the channel bed.

The magnitude, duration, and post-flood flow regime of future beach/habitat-building flows are currently under debate (Argonne National Laboratory, 1999; Melis, 1998). One proposal involves releases of up to 60,000 cfs for several days, followed by fluctuating (load-following) flows. Discussion of this proposal focused on issues of hypothesis testing and multiple treatments (Sit and Taylor, 1998). This is an appropriate discussion within the framework of designing adaptive management experiments and focuses on the appropriate magnitude and duration of beach/habitat-building flows, the sequence of experimental floods most likely to demonstrate clear results, and the utility of a fluctuating post-flood regime for conserving deposited sand.

Research during and after the 1996 controlled flood suggests that a shorter-duration beach/habitat-building flow of larger magnitude may be more effective than the flood in 1996 (Schmidt, 1999). The concentration of sediment in suspension decreased during the 1996 flood, indicating that channel sediments available for redistribution decreased over its course (Smith, 1999; Topping et al., 1999). Bar deposition rates were larger, while suspended sand concentrations were higher early in the flood (Andrews et al., 1999; Schmidt, 1999), a result supported by numerical simulations of the flow and transport field (Wiele et al., 1999). A shorter-duration beach/habitat-building flow is also supported by observations that most debris-fan reworking occurred during the initial hours of the controlled flood (AGU, 1999).

Numerical modeling of the flow and transport field provides a means of evaluating effects of different management options and a means of forecasting conditions at locations where monitoring is not conducted. Both the 1997 and 1998 strategic plans emphasize the utility of numerical modeling and incorporate it as part of the long-term monitoring program. Both plans also emphasize the desirability of developing remote sensing methods for basic water and sediment monitoring, and the Physical Resources Program is actively exploring less invasive means of collecting adequate monitoring data. The present state of the art in both numerical modeling and remote sensing, however, is such that on-the-ground long-term monitoring and periodic detailed measurements of local processes are still required.

The Physical Resources Program was reviewed by a protocol evaluation program panel in August 1998 (Wohl et al., 1998). The panel noted that the program is well managed and integrated with an admirable degree of cooperation among investigators. Most ongoing projects received unqualified panel support. The review was wide-ranging and frank, with extensive and open cooperation by Center staff and program investigators, providing a strong example for comparable reviews in the Center's other resource programs.

Weaknesses and Alternative Approaches

The 1998 Strategic Plan identifies fundamental physical science questions that remain to be answered; however, their relative importance and connections are lost within a text that wanders among abstract monitoring goals, strategies, and external review guidelines not unique to the Physical Resources Program. To be an effective planning document, the program description needs to clearly and directly present its accomplishments, goals, and overall strategy, following an organization that parallels that of the other program descriptions.

An immediate goal is to complete a sediment budget with acceptable levels of accuracy for all components. Proposed reductions in the program budget in fiscal year 2000 and beyond may hamper this effort, delaying implementation of the long-term sediment monitoring program and impacting ecosystem experiments. A robust and accurate sediment budget is required for testing hypotheses linking ecosystem responses to dam operations. Particularly important is completion of a sand budget for Marble Canyon, including the long-term trend in storage, spatial and temporal variability in storage, and residence time of tributary-derived sediment. Sand storage in this reach may be reduced to a level that will not sustain valued recreational and ecological resources.

Management experiments designed to evaluate the effectiveness of different beach/habitat-building flows face institutional and legal constraints regarding both the magnitude and timing of the flows. Resource impacts of beach/habitat-building flows have been examined only for floods below 45,000 cfs (Ralston et al., 1998). As discussed previously (both in the Glen Canyon Dam Environmental Impact Statement and in planning before the 1996 controlled flood), this evaluation should be expanded to include a much wider range of discharge (e.g., to 90,000 cfs) so that this information is available for informed management decisions.

An additional concern is the role of flood timing in preserving sand in reaches closest to the dam. Changes in concentration and grain size of sand transported during the 1996 beach/habitat-building flow suggested that sediment available for transport was reduced over the course of the flood (Rubin et al., 1998; Topping et al., 1999). This suggests that beach/habitat-building flows for conserving sand and building beaches may be most profitably scheduled shortly after tributary floods, when the amount of sediment available in the channel is at a maximum. This is particularly the case in Marble Canyon, where the post-dam decrease in sediment supply is largest and where long-term availability of camping beaches and riparian habitat depends almost entirely on sediment inputs from the Paria River. Discussions on the timing of the 1996 controlled flood included the possibility of an October flood, and the Glen Canyon Dam Environmental Impact Statement suggested that beach/habitat-building flows could be timed to follow tributary floods in the late summer. Subsequent analyses, however, have focused entirely on January through June. Evaluation of potential sediment conservation benefits and resource effects should be extended to other months of the year.

Mobility of large boulders and cobbles in rapids during beach/habitat-building flows also requires further investigation. Although the response of debris fans to flows is mentioned in the Strategic Plan, there appears to be no funding for continued studies of such responses. An important objective of experimental flows is to redistribute coarse grains on debris fans and maintain runnable rapids. Observations of boulder entrainment during beach/habitat-building flows of different magnitudes are needed to develop an ability to forecast their effectiveness at maintaining rapids.

Much of the work on flow and sediment has been conducted by the U.S. Geological Survey through large, multiproject contracts. This contracting method has the effect of putting a management layer between the Center and individual principal investigators, making it difficult for Center staff to hold individual principal investigators accountable, and it increases chances for murky communication. Although less than in the past, this buffering is still the case for fiscal year 1998–1999 funding. An important example occurred in September 1998. Over the course of one week, two large floods on the Paria River discharged into the mainstem a volume of sediment larger than the Paria River's mean annual load (D. J. Topping, U.S. Geological Survey, personal communication, 1998). This presented an opportunity to track the transport and storage of a large sediment input, an important and poorly understood factor for evaluating timing of beach/habitat-building flows for sediment conservation, particularly for Marble Canyon. Although the U.S. Geological Survey was to sample sediment transport at two downstream sites during this period, the planned sampling did not occur. Although steps have been taken to reduce chances of another missed opportunity, part of the problem lies in the physical and institutional separation between the Center and the U.S. Geological Survey and the Arizona Geological Survey staff assigned to conduct the monitoring. A more suitable long-term solution would be to give the Center more direct control of monitoring, so that the required data are consistently collected.

Biological Resources Program

The biological resources section of the Strategic Plan presents important ideas about the value of an ecosystem paradigm and monitoring principles, but the program itself is narrow, even when viewed collectively. Legal and institutional requirements mandate studies of only a few "key" species (e.g., humpback chub), although the examination of other ecosystem components is critical to understanding the roles of the few species emphasized. The closing and subsequent operation of Glen Canyon Dam have had tremendous repercussions on the native biota. Although implicit in documents produced by Glen Canyon Environmental Studies and other documents produced later, few documents attempt overall synthesis of these effects (exceptions include, for example, Valdez and Carothers, 1998; Patten, 1998). They may thus not yet be fully appreciated by all the parties concerned. The 1998 Strategic Plan briefly mentions pre-dam and post-dam conditions and time scales for the research program. Lack of long-term historical synthesis hampers objective evaluation of the natural state of aquatic biota, as well as the establishment of "baseline" (approximately pre-dam) conditions. Perhaps it is assumed that narrative (and in some cases quantitative) syntheses for different ecosystem components in the Grand Canyon (e.g., general: Carothers and Aitchison, 1976, Johnson, 1977; vegetation: Johnson, 1991, Turner and Karpiscak, 1980, Webb, 1996; fishes: Carothers and Minckley, 1981, Minckley, 1991, Suttkus and Clemmer, 1979; birds: Brown et al., 1987; mammals: Hoffmeister, 1971, Ruffner et al., 1978), other parts of the Colorado River (general: Ohmart et al., 1988 and citations; vegetation: Anderson and Ohmart, 1985 and citations; birds: Rosenberg et al., 1991;

fishes: Minckley 1979, 1985), and elsewhere (vegetation: Brown, 1994, Hastings and Turner, 1966) are generally known, but this is hardly the case.

Glen Canyon Dam resulted in complex physical, chemical, and biotic impacts on biological resources in the Grand Canyon. The river corridor biota were subjected to sudden, general stabilization of essentially all variables to which they were most likely adapted. Major impacts occurred in volumes and patterns of flow and temperatures, along with altered quantity and quality of sediments, including organic materials. Chemical variations downstream, ionic composition including nutrient quantities and qualities, and dissolved organics were buffered and otherwise modified in Lake Powell. The scenario of change further involved biotic impacts amplified by direct and indirect species and community interactions as the ecosystem shifted in response to novel nutrient supply, patterns of flow, seasonality, and turbidity. Native survivors were joined by nonnative colonists, some of them having been there before the dam and others having been dispersed from elsewhere. Colonization was augmented through stocking diverse invertebrates to establish a food base for recreational fisheries.

Over time, an interacting, ever-changing species pool has resulted in the biological communities existing today, with nonnative species living in a new environment(s) along with a few surviving natives. With the installation of temperature control structures currently proposed at Glen Canyon Dam (U.S. Bureau of Reclamation, 1999), the ecosystem will again be altered. Both the biota and the researchers studying it are thus confronted with a continuous "moving target," resulting in part in the complexity and confusion evident in parts of the Biological Resources Program.

Synthesis of Previous Knowledge

As noted elsewhere, the revised Strategic Plan provides modest evidence of synthesis of existing knowledge in describing research and monitoring; this is especially true for the Biological Resources Program. Much of the section on this program relates to broad, philosophical principles of research and monitoring, presented much as a textbook on ecosystem management or ecology and providing few specific indications of how it all relates to the Grand Canyon ecosystem.

Evaluation and use of past research knowledge is, however, becoming part of the program. The most comprehensive synthesis of information appearing to date has been that of Valdez and Carothers (1998), produced as part a previous U.S. Bureau of Reclamation contract. Other information (Patten, 1998) has been developed through voluntary efforts by researchers funded in the past by the Glen Canyon Environmental Studies and in some cases by the Center (e.g., Douglas and Marsh, 1996, 1998; Marsh and Douglas, 1997; Marzolf et al., 1998; Schmidt et al., 1998). Other information appeared in the AGU (1999) volume on the 1996 controlled flood.

Relationships of some biological resources to physical features influenced by the Glen Canyon Dam, such as temperature, fluctuating flow, and some patterns of sedimentation (e.g., importance of "backwaters" as nursery areas for fishes) are well enough understood for formulation and testing of hypotheses relative to reproduction and recruitment of fishes. Information on the nature and interrelations of other features of the physicochemical setting, based on conditions introduced to tailwaters, and downstream on the presence and operation of the dam, provides a framework for formulating and testing hypotheses on controls and mechanisms of response of other biological components of the ecosystem, as well. Linkages between and among biological and various other resource categories remain poorly articulated. It is therefore critical that the Biological Resources Program be closely integrated within itself and that, at a minimum, the Biological Resources and Physical Resources programs be tightly interwoven; it is not yet apparent that either condition is satisfied.

Likely Effectiveness of the Strategic Plan

There is strong evidence that Center staff are actively seeking to identify issues and criteria for general biological monitoring, and the search has been fruitful for some components. Food base analyses are far enough along for formulation and testing of hypotheses, as are some aspects of on-ground monitoring of waterfowl and breeding birds, and remote sensing of such things as riparian vegetative communities (the latter are commendably integrated with physical resources such as sandbars). Further, some program components focusing on meeting the compliance and impact assessment requirements of the Glen Canyon Dam

Environmental Impact Statement, the Record of Decision, and alternatives to avoid negative biological opinions show likelihood of success. It is also significant that, along with the Physical Resources Program, requirements anticipated for the Biological Resources Program clearly played a pivotal role in commissioning the conceptual modeling exercise. As noted, the committee considers conceptual modeling to be a major accomplishment, pointing toward an ecosystem paradigm for the Grand Canyon.

Despite these positive aspects, some important parts of the Biological Resources Program show little evidence of being based upon an ecosystem paradigm and may thus prove inadequate for developing adaptive management strategies. This is most evident when management objectives and information needs, as well as currently funded projects for fishes, are considered. Most emphasis is clearly upon: (1) recreational fisheries, (2) avoidance of jeopardy opinion under the Endangered Species Act for endangered humpback chub, and (3) enhancement and/or maintenance to assure compliance with recovery stipulations and to preclude future listing or future jeopardy opinions for other candidate species (flannelmouth sucker) or listed species (razorback sucker, Kanab ambersnail, Southwest Willow Flycatcher). Biological research on these components has been and remains driven by actual and perceived needs to satisfy reasonable and prudent alternatives and other mandates and agreements rather than by needs for ecosystem management.

Until linkages are defined among various biological resource components, single-species questions and accumulation of species-specific ecological information will prevail. Studies of fishes should include, for example, comparisons between quantities and qualities of foods (calories) acquired from different sources, such as aquatic vs. terrestrial. At the community level, comparisons of food habits between tributary vs. mainstem, or up- vs. downstream, would be informative. Inferences from food supply and demand could be expanded to other biological features, such as rates and patterns of growth and reproduction within different habitats. Annual, seasonal, and daily movements might further be examined with an eye toward defining transience vs. permanence of fish community structure. Ongoing and anticipated studies of aquatic food base, terrestrial vegetation, and terrestrial vertebrates tend to have more of an ecosystem flavor. They are, however, apparently assigned lower priorities in the program (other than for sport fishes and the endangered Southwestern Willow Flycatcher, each of which comprises a single-species initiative with strong ecosystem overtones).

Weaknesses and Alternative Approaches -

Although studies currently underway may contribute to goals for sport fishes and listed species and thus continue to add to existing databases on these few biotic ecosystem components, their integrative contribution to ecosystem understanding may prove minimal. This committee anticipates that findings and discussions from the 1998 conceptual modeling workshop may further reveal several serious experimental design problems in earlier and ongoing biological research and management in the Grand Canyon.

Glen Canyon Environmental Studies Phase II was criticized for the lack of cohesiveness of its research program, caused in large part by the unanticipated environmental impact statement preparation requirements within a research effort already underway. There seems a comparable danger in the demands on the Center to conduct, administer, and/or coordinate compliance requirements of biological opinions, programmatic agreements, and environmental assessments. Of 80 biological information needs in the 1998 Strategic Plan, more than 40 descend directly from requirements mandated by federal listing or candidacy of individual species. Only about seven information needs, or combination of needs, in the plans seem definitely positioned within an ecosystem paradigm. These are listed in Table 4.1, along with clarifications in brackets. The Biological Resources Program should be reconstructed with hypotheses directed toward anticipated needs for adaptive management of the *system* as the support engine for its biotic components, rather than for managing the components as impacted by operations of Glen Canyon Dam.

It seems clear this is understood and accepted by some individuals and agencies involved in the Adaptive Management Program. Others, however, either fail to understand or do not embrace the ecosystem paradigm, or are unwilling to recognize that the Grand Canyon ecosystem as a whole should ultimately be the unit of management. This results in apparent confusion, resulting in stop-gap attempts to salvage what remains of the native biota, whatever it might be (or might have been). An overall desired state toward which management may be directed has yet to be defined. An ecosystem vision, the lack of which is discussed elsewhere, clearly needs to be developed within mandated constraints before adaptive management can be fulfilled. This committee notes that a key goal should be to move the Adaptive Management Program from an exercise in

TABLE 4.1 – Biological Resource Program Information Needs

Information needs (of a total of 80) quoted or paraphrased from the Center's 1998 Strategic Plan, that stand alone or in combination with others (indicated by boldfaced, Roman numerals **I** through **VII**) to fall within an ecosystem paradigm for the Grand Canyon ecosystem.

(I) IN 1.1, "Determine status and trends in...food base species composition and population structure, density and distribution and the influence of ecologically significant processes."

(II) IN 1.2, "Determine the effects of past, present, and future dam operations under the approved operations criteria on the aquatic food base species composition, population structure, density, and distribution..."

(III) IN 2.1, "Determine ecosystem requirements, population character, and structure [required] to maintain naturally reproducing populations [of trout] ..."

(IV) IN 2.7, "Determine the trophic relationship between trout and the aquatic food base including the size of...food base required to sustain the desired trout population [and impacts of trout on the food base relative to downstream system requirements]."

(V) IN 3/4.7, "Determine origins of fish food resources, energy pathways, and nutrient sources important to their production, and the effects of Glen Canyon Dam operations on these resources...Evaluate linkages between the aquatic food base and health and sustainability of HBC [= humpback chub; replace with 'native fish'] populations."

(VI) IN 11.1 and 12.1 combined, respectively, "Define and specify ecology of native [terrestrial] faunal components, especially threatened and endangered species; including evolutionary and environmental changes, natural range of variations, linkages, interdependencies, and requirements." And, "Identify...species potentially affected by dam operations and determine effects on distribution, abundance, and population structure."

continues

TABLE 4.1 Continued

(VII) IN 16.1 and IN 16.4 combined, respectively, "Determine distribution and abundance of native and non-native riparian and upland vegetation, including federal-, state- and tribal-listed sensitive species, old high water zone, new high water zone, and nearshore marshes." And "Determine the effects of current and proposed dam operations [on plant communities]..."

SOURCE: Center (1998).

impact assessment toward ecosystem management.

It has long been recognized (Clarkson et al., 1994) that cold, hypolimnic water releases from Glen Canyon Dam have overwhelming impacts on aquatic biota of the Grand Canyon ecosystem. Thus, a lack of attention in the Strategic Plan to temperature control as a potential eco-system manipulation (other than as a generally worded information need) is inappropriate and should be reversed. Second to temperature are impacts of nonnative fishes that prey upon and compete with native species (Minckley, 1991). These two forms of environmental resistance overlap in a justifiable concern that increased water temperature below Glen Canyon Dam will enhance populations of nonnative competitors and predators as well as native, warm-water species. Assessment of such expected and predictable interactions should take a high priority in adaptive management considerations as well, as is addressed from the operational viewpoint in the U.S. Bureau of Reclamation's environmental assessment (1999) on retrofitting Glen Canyon Dam with temperature-control devices.

This committee recommends the Biological Resources Program be reconstituted into two broadly overlapping elements. A first should clearly emphasize testing of hypotheses and implementation of manage-ment actions to further compliance with management objectives related to the Endangered Species Act, the Glen Canyon Dam Environmental Impact Statement, and other agreements. The second element should be dedicated to embracing the ecosystem as a whole, which is crucial both for explaining individual and interacting resource effects and for fulfilling the intent of the Grand Canyon Protection Act. The major goal of this second element should pertain to maintaining ecosystem function at levels defined by historical reconstructions, tempered by realistic consideration of the constraints of human uses.

Because listed species and other organisms of concern are important parts of the ecosystem, efforts under the second element should strive insofar as possible to incorporate them into the testing of hypotheses and implementation of actions at the ecosystem level. The Adaptive Management Program should then be designed by bonding general ecosystem concerns with those for species of special emphasis to ensure sustainability near their defined levels. This committee suggests the following actions under this recommendation:

• Efforts in the Biological Resources Program should be refocused at the community/ecosystem level, segregating yet accommodating various subunits (species or other components) of both ecological and social importance in a hierarchical manner (O'Neill et al., 1986).

• A succinct historical synthesis should be commissioned, describing natural ecological conditions based on qualitative and quantitative (when possible) information in the literature (see, for example, Brown et al., 1987; Clarkson et al., 1994; NRC, 1991; Stevens, 1983; Turner and Karpiscak, 1980; Webb, 1996). This should be accompanied by a qualitative and quantitative systematic assessment of the individual and collective effects of dam emplacement and operations. A comparable, parallel, authoritative history should be prepared for legal and political agreements requiring environmental compliance, including assessment of their individual and collective influences on prior research and management actions (see, for example, Carothers and House, 1996; Marzolf, 1991) and how they continue to influence the Center's functions.

• Management objectives and information needs for the Biological Resources Program should be generalized, condensed, and stated explicitly as falsifiable hypotheses, realigned within one of the two elements of emphasis. Research toward answering questions and management actions to maintain ecosystem sustainability should, whenever practical, incorporate those required for compliance with political or legal requirements.

• Major features to be studied from the ecosystem perspective should be explicitly defined and placed in the context of the physical, biological, cultural, and socioeconomic programs. This compilation of ecological priorities would contribute to a decision-making process, weighted as objectively as possible, for evaluating alternative recommendations to the Secretary of the Interior.

• These suggestions are not exclusive to the Biological Resources Program. All Grand Canyon resources should be considered and integrated as historical documentation is prepared, major ecosystem components identified, and research and management proceeds. This provides an opportunity that, with sufficient emphasis, can contribute significantly to highly desirable, across-program integration and alternatives analysis.

Sociocultural Resources Program

The 1998 Strategic Plan combines cultural resources, including tribal programs, and socioeconomic resources under a single program. Of the Center's resource programs, the revised plan for sociocultural resources is most explicitly structured to indicate how proposed research and monitoring activities address specific information needs that address, in turn, current management objectives. Two of the three 1998 cultural resources research grants involve physical science components that assess the archaeological effects of dam operations and thus reflect a growing level of integration across programs. Progress has also been made toward coordinating the Center's Cultural Resources Program with the U.S. Bureau of Reclamation and U.S. National Park Service Programmatic Agreement with the tribes (Dongoske and Yeatts, 1998).

These developments are promising in several respects. In principle, combining cultural and socioeconomic programs would facilitate comparison of effects of dam operations on different social groups. The Center correctly recognizes that it is a mistake to treat tribal interests as exclusively "cultural" and nontribal interests as exclusively "socioeconomic." In addition, interests vary within and across groups, and they include complex combinations of conservation, preservation, and economic development interests. Among the more important and least understood issues for social research are the following: *what* resource effects are valued by different groups, *how* they are experienced and valued, and *how much* they are valued. Previous research within the Glen Canyon Environmental Studies addressed the "what" and "how much" questions, with less formal research on identifying common ground and basic differences, or on changes in "how" downstream resources and resource effects are experienced and valued (cf. Smith, 1998). Communication across the cultural, socioeconomic, and other research programs could shed light on these issues.

Underneath the sociocultural umbrella, the Cultural Resources and Socioeconomic Resources programs are still presented as separate programs in the 1998 Strategic Plan, so they are treated separately below. As a general concern about staffing the newly combined sociocultural program, however, it should be stressed that one full-time employee to serve both the cultural and socioeconomic programs is inadequate because of both work load considerations and the range of disciplines and level of training required to manage these two programs. Employing only one full-time staff member to manage the two programs would likely lead to ineffectiveness in both programs.

Cultural Resources Program

The Cultural Resources Program is the third largest Center program after the Biological Resources and Physical Resources programs, and it is far larger than the Socioeconomic Resources Program. It also has the most complex organizational structure. Its main components are:

- Cultural resources monitoring and research
- Cooperative tribal projects
- Individual tribal projects

The monitoring and research program addresses management objectives and information needs identified by stakeholders, which established the structure of the 1998 Strategic Plan. Current management objectives focus on monitoring and protection of archaeological sites.

The tribes have a sovereign status, and the federal government has a trust responsibility toward them, which necessitates some distinct tribal programs (Tsosie, 1998). The Center has recently compiled Glen Canyon Environmental Studies Phase II synthesis reports on tribal interests in and perspectives on Grand Canyon resources for five of the six tribes that are involved; those reports provide a wealth of insight and information that has broad value for other science and stakeholder groups (Ferguson, 1998; Hart, 1995; Phillips and Jackson, 1997; Roberts et al., 1995; Stoffle et al.,

1994, 1995). The Center's Plan is also sensitive to the need for confidentiality for some tribal cultural resources information. Individual tribal programs support tribal monitoring and research interests. They provide for full tribal involvement in the identification, design, and completion of the research. They "may investigate resources that have cultural values to Native Americans but are outside western notions of cultural resources" (Center, 1998, p. 99). Cooperative programs emphasize education, training, and information dissemination projects with tribal groups.

These Center programs represent continuation of the trend that began in 1990 toward greater tribal involvement in cultural resources programs associated with dam operations. The Center's main challenge will be to coordinate and integrate these activities, both logistically and intellectually. If the Center's efforts are successful, the Center's program could serve as a partial model for working with other stakeholder cultural groups interested in participatory research, education, and conservation.

In addition to Center and tribal programs, there is a separate Programmatic Agreement among the U.S. Bureau of Reclamation, the U.S. National Park Service, the Advisory Council on Historic Preservation, and seven Tribes (the Havasupai and San Juan Southern Paiute had not signed as of April 1999) to monitor and mitigate dam-operation impacts on cultural resources eligible for listing as historic properties. The Agreement's geographic scope has extended laterally to include surveys of the 256,000 cfs flood level, which roughly encompasses the 100-year flood recurrence interval (T. Melis, Grand Canyon Monitoring and Research Center, personal communication, 1999).

The Center has a broader mandate than the Programmatic Agreement to assess the effects of dam operations on downstream cultural resources, including "archaeological, ethnographic, ethno-botanical, faunal, and physical resources" (Center, 1997), whether or not they are eligible for listing as historic properties. Unlike the Programmatic Agreement, however, the Center is not required to mitigate those impacts. In an effort to coordinate the Center's Cultural Resources Program with the Programmatic Agreement, the Center was asked to administer both programs in 1997–1998. This arrangement proved unwieldy because the Bureau of Reclamation and the National Park Service have legal responsibility for implementing the Programmatic Agreement, which cannot be delegated to the Center. This resulted in delays and procedural complications. The Bureau of Reclamation therefore resumed direct administration of the Programmatic Agreement in fiscal year 1998.

Dongoskc and Ycatts (1998) dcvclopcd a plan to bcttcr coordinate the two programs, which was adopted by the Technical Work Group.

During the first research cycle, the Cultural Resources Program let the following three research grant contracts:

1. Test and apply a geomorphic model related to erosion of pre-dam river terraces in the Colorado River ecosystem containing cultural materials. Awarded to SWCA, Inc.

2. Model mainstem flow and sediment dynamics at selected cultural resource locations. Awarded to the U.S. Geological Survey.

3. A cultural resources synthesis project to draw together Glen Canyon Environmental Studies and related research. Awarded to SWCA, Inc.

The first two projects indicate close coordination with physical resources monitoring and research and clearly examine the effects of flow regimes on archaeological site erosion. The third project addresses the need for synthesis and integration of previous cultural resources research.

Synthesis of Previous Knowledge   The 1998 Strategic Plan provides a clear synopsis of past research, environmental impact statements, Programmatic Agreement research, and new Center studies. The Center has begun an important synthesis of these previous cultural studies in the Grand Canyon and of data assembled under them (SWCA, 1998). A previous review of archaeological site information had been prepared with support from the Glen Canyon Environmental Studies (Fairley et al., 1994).

The Glen Canyon Environmental Studies also previously commissioned broad assessments of Grand Canyon resources by tribes and tribal consortia. These include:

1. **Havasupai** — Not currently participating.

2. **Hopi** — Ferguson, T. J. 1998. *Ongtupqa niqw Pisisvayu (Salt Canyon and the Colorado River). The Hopi People and the Grand Canyon.* Produced by the Hopi Cultural Preservation Office, under the guidance of the Hopi Cultural Resources Advisory Task Team, and under contract with the U.S. Bureau

of Reclamation. Tucson, Ariz.: Anthropological Research.

3. **Hualapai** — Phillips, A. M., III and L. Jackson. December 31, 1997. Monitoring Hualapai ethnobotanical resources along the Colorado River, 1997. Annual Report. Hualapai Tribe, Cultural Resources Division.

4. **Navajo** — Roberts, A., R. M. Begay, and K. B. Kelley. August 9, 1995. *Bits'iis Nineezi (The River of Neverending Life): Navajo History and Cultural Resources of the Grand Canyon and the Colorado River.* Window Rock, Ariz.: Navajo Nation Historic Preservation Department.

5. **Southern Paiute Consortium** — (1) Stoffle, R. W. et al. September 1995. *Itus, Auv, Te'ek (Past, Present, Future). Managing Southern Paiute resources in the Colorado River Corridor.* Pipe Spring, Ariz.: Southern Paiute Consortium, and Tucson Bureau of Applied Research in Anthropology, University of Arizona, (2) Stoffle, R. W. et al. 1994. *Piapaxa 'Uipi (Big River Canyon).* Tucson, Ariz.: Bureau of Applied Research in Anthropology, University of Arizona.

6. **Zuni** — Hart, E. R. July 21, 1995. Zuni and the Grand Canyon: A Glen Canyon Environmental Studies Report. Zuni GCES Ethnohistorical Report. Seattle, Wash.: Institute of the North American West.

These reports and related publications shed light on the relationships between Grand Canyon "resources" and "values," a theme central to the Adaptive Management Program (cf. Bravo and Susanyatame, 1997; Dongoske, 1996; Kelley and Francis, 1994; and the *SAA Bulletin* "Working Together" series, 1993—). They present a range of ways for articulating and understanding experiences, uses, and concerns in the Grand Canyon.

The current synthesis project appears very capable of incorporating previous research on archaeological and tribal resources, which would be a major accomplishment. However, the Center's cultural resources synthesis has yet to encompass all cultural groups or to envision a dialogue among the concerns and views of different groups. The Center took an important step in this direction by convening a cultural resources workshop for the March 1999 Technical Work Group meeting to present and discuss current research and synthesis projects. Regular workshops of this sort could help illuminate the cultural bases of adaptive management. To broaden the scope of the Cultural Resources Program, the Center might draw upon historical and contemporary studies by and about explorers, travelers, prospectors, developers, river runners, dam operators, environmentalists, and scientists in the Canyon (e.g., Lavender, 1985; Morehouse, 1996; Powell, 1874; Riebsame, 1997; Webb, 1996).

Likely Effectiveness of the Strategic Plan   The Grand Canyon Protection Act, the Glen Canyon Dam Environmental Impact Statement, and the Record of Decision all stress the importance of cultural resources protection and consultation with tribes. The Glen Canyon Dam Environmental Impact Statement examines cultural resources in sections of the "Affected Environment" and "Environmental Consequences" chapters. The Glen Canyon Dam Environmental Impact Statement treats cultural resources as either archaeological sites or traditional cultural properties, but it does not specifically discuss tribal concerns about other natural resources and socioeconomic issues.

Of all the Center programs, the strategic plan for cultural resources most closely follows management objectives and information needs identified by stakeholders. In that respect, it seems highly responsive to the new Adaptive Management Program. This approach, however, raises some concerns. If management objectives and information needs are revised annually or in dramatic ways, a Strategic Plan based exclusively upon them could become obsolete. If management objectives and information needs are poorly coordinated, as is the case *across* major sociocultural resource categories (i.e., cultural resources, recreation, water, hydropower, etc.), the program would lose coherence. And if management objectives and information needs are missing, the program has no way of identifying them for consideration by stakeholders (on this point, see the socioeconomic resources section below).

Because tribes and other social groups are differentially involved in the Adaptive Management Program, the Center may become more responsive to those that take a greater role in the process. This issue should be anticipated in the Strategic Plan by considering ways to maintain contact with and involve of all tribes and groups.

The broader role of cultural resources in adaptive management, comparable with the role of ecosystem science, has perhaps not yet been fully envisioned. To develop a broader perspective, the Center might find it useful to consider previous research in the fields of cultural ecology (e.g., Bennett, 1969; Butzer, 1989; Denevan, 1983; Ellen, 1982), which has developed theories of adaptation, adaptive management, and adaptive strategies; global environmental change (May, 1996; Smith, 1997; Smithers and Smit, 1997); environmental philosophy (Griffiths, 1996; Light and Katz, 1996); and the emerging field of cultural studies, all of which explore different aspects of human adaptation.

Weaknesses and Alternative Approaches Two separate objectives regarding Grand Canyon archaeology should be integrated for the more effective realization of both the immediate goal of locating, monitoring, and protecting, and the long-range goal of interpreting and understanding. Sites and isolated remains to a large degree reflect the physical and biological state of the Grand Canyon ecosystem in prehistoric times. Physical and biological studies should also include efforts to describe past environmental states (historical studies) and to identify current changes in variables influenced by prior patterns of human occupation. A model or chronological series of models of land use and settlement pattern in riparian zones should be developed. With such models, the current physical and biological studies could contribute more to an understanding of human occupations than could be achieved by focusing solely on mechanistic processes of site destruction and preservation. Undiscovered isolated remains and sites can be anticipated, protected, and interpreted in light of models that relate them to environmental variables in riverine areas.

Tribal perspectives on resources affected by dam operations are a source of valuable insights into the physical and biological parameters affecting prehistoric occupations along the river. When site effects are mitigated, there is an opportunity to compare archaeological evidence for resource relationships with ethnographic accounts. Parallels and differences would be of significance to both archaeologists and tribal members.

How can the information from tribal reports and perspectives be integrated with other aspects of ecosystem monitoring, research, and modeling? A first step is encouraging the tribes to articulate their own concepts of ecosystem and its important components, particularly with regard to river and riparian zones. To the extent that they identify indicators of ecosystem integrity from their perspectives, building upon the synthesis reports produced to date, the Center and other scientific monitoring programs could incorporate these variables and provide useful information in addition to that available from tribal programs in the Grand Canyon. Even though tribal concepts of ecosystems may not be the same as those of currently practiced science, points of productive intersection can be sought.

Consultation with tribes about their identification of critical ecosystem variables seems to be an urgent step. Dam operations and related changes have immediate impacts on those living in and spiritually associated with the Grand Canyon. Conceptual modeling should certainly address the interaction of ecosystem components and ecosystem integrity with respect to tribal social and economic activities and values. If this important issue has been addressed, it is not apparent in the literature provided. The key problem with present cultural resource management objectives in the strategic plan is a lack of integration—integration between the ethnographic and archaeological programs and between these programs and the ecosystem management paradigm.

It is a matter of continuing concern that the Havasupai tribe has not joined the monitoring and research program. The Center has contacted the tribe and will presumably continue to contact them in order to will fulfill its trust and scientific responsibilities, but the tribe's decision not to participate must be respected. Also, as discussed below, effective participation of the all tribes depends upon the resolution of a number of key financial and programmatic issues.

The committee is concerned about reduced funding for tribal participation in the Adaptive Management Program. Tribal participation did not receive early support in the Glen Canyon Environmental Studies, but it grew and contributed in important ways in the 1990s and, for reasons indicated above, should expand rather than contract (NRC, 1996a). As a federal program, the Adaptive Management Program has trust responsibilities to the tribes. The Center's Strategic Plan displays sensitivity to those responsibilities and it correctly focuses on tribal participation in monitoring and research. Any reallocation of resources that diminished participation in monitoring and research activities would aggravate the general trend away from Grand Canyon investigations. This committee recommends that resources be secured for full tribal participation in all aspects of monitoring, research, and communication in the

Adaptive Management Program, without reducing other components of the cultural resources monitoring and research program.

Socioeconomic Resources Program

Discussion of the socioeconomic dimensions of the Strategic Plan differs from the discussion of its other components because the Plan provides little to evaluate. This section outlines what is missing and explains why it matters.

Before describing the Strategic Plan's Socioeconomics Resources Program, it should be pointed out that there are many aspects of "socio-economics," including environmental economics, geography, historical studies, institutional and policy analysis, and recreational sociology. The Center is currently supporting important research on recreational socio-logy in the Grand Canyon, and this committee has recommended that historical and institutional studies be conducted as part of the broader Adaptive Management Program. This section of the report focuses on the major resource area that was included in the Glen Canyon Dam Environmental Impact Statement and previous National Research Council reviews, but is not adequately incorporated within the Center's resource programs: economic values of downstream resources in the Grand Canyon. Given the importance of this topic for analyzing the effects of dam operations and for formulating recommendations to the Secretary of the Interior, this committee is concerned that it may not be strengthened. Since the last National Research Council report, there has been a great deal of research conducted on these and related topics in natural resource valuation. Advances in environmental economics and some recent studies are described in Appendix F.

The Strategic Plan does not adequately explore the possibility that some common insights in environmental economics might be exploited to clarify the process of advising and policy-making in the Grand Canyon. The last National Research Council report concerning the Grand Canyon provided a thorough and careful review of the issues involved (NRC, 1996a); it is not necessary to reiterate that review here. Efforts of the prior National Research Council committee to explain the full scope of the "economic" dimensions of Grand Canyon management do not, however, seem to have made much of a difference regarding the Center's approaches to these matters. The Center does not presently have any in-house expertise in environmental economics, which may account for this oversight.

Why is some specialized environmental economics expertise necessary? Economics is about the allocation of scarce resources among competing end uses. Any adjustment to dam operations is likely to affect Grand Canyon resources. If the effects on these resources are beneficial from the perspective of all stakeholders, then the adjustment is likely to be uncontroversial. This would imply an unambiguous improvement in the "common good," and the change would likely be made. Conversely, if stakeholders universally perceive the effects of a change as negative, then the change would likely not occur, as the status quo would then be preferred by everyone. Management decisions can be difficult, however, when an adjustment would result in winners and losers. In these cases, managers must weigh the gains to the winners against the losses to the losers. If circumstances dictate that each type of stakeholder be given equal weight in the decision process, and if the winners' gains exceed the losers' losses, the change should be implemented. Often, however, the distributional consequences of a proposed change are important, and understanding the individual magnitudes of these gains and losses is only the first step in the decision-making process.

Synthesis of Previous Knowledge A large amount of research conducted since the last National Research Council review in the mid-1990s bears on current and future efforts to establish the relative social values of competing objectives in the management of Grand Canyon resources (cf. Appendix F of this report. For an introductory discussion, see Callan and Thomas, 1996; see Hanley et al., 1997 for an intermediate treatment). Market costs and benefits are relatively easy to measure and track because they are captured by changes in prices and costs. The market component of costs and benefits is relatively uncontroversial and appears to have been accounted for adequately in the Center's work. The Center, however, does not have an economist who follows the literature on methodologies and applications concerning the valuation of "nonmarket" environmental goods. This may explain why, in Center documents, "economic" issues repeatedly devolve to a subset consisting primarily of hydropower costs and "regional economic impacts" in the form of revenues of recreational guides and outfitters.

Likely Effectiveness of the Strategic Plan The current Strategic Plan ignores all but a very restricted subset of the economic issues pertinent to Grand Canyon management and is thus unlikely to be "effective." A small set of easy-to-measure economic quantities has been targeted for attention, but these do not constitute the full set of relevant economic quantities, nor necessarily the most appropriate ones.

In some cases, it is relatively easy to assign estimates of the costs of a management decision. Where a proposed change in management will affect the prices paid by consumers of hydroelectric power, for example, there are standard methods to determine the relevant social costs. These techniques are relatively straightforward and uncontroversial and can be estimated from observed historical market data. Also, these private cost estimates are likely to be available, because the relevant group of stakeholders is typically well organized and is adequately funded to conduct the research necessary to generate defensible cost information.

It is often much harder to generate equally defensible estimates of the social benefits associated with an environmental management decision. The only direct market information associated with conditions in the Grand Canyon ecosystem might involve estimates of the total revenues of guides and outfitters serving recreational users, along with some estimated number of local jobs attributable to the existence of these resources. Economic theory is, however, clear on the fact that revenues of collateral business activity do not represent a full measure of social value of the existence of the resource, let alone the change in social values associated with variations in the resource's condition. At best, regional economic activity effects are a measure of the distributional consequences of some change, not the overall benefits to society of that change. The correct measure, roughly speaking, is the excess of "willingness to pay" over what people actually pay to enjoy the ecological and recreational services of Grand Canyon resources. Shifts in demands for these resources as a result of changes in their management will alter this measure of social value.

Understanding the social benefits associated with improved ecological or recreational conditions in the Grand Canyon requires information about society's willingness to pay for enhancement of ecological conditions or for better recreational opportunities. The problem stems from the fact that, unlike the case for valuing hydropower market consequences, these things are not traded at explicit prices in traditional markets. Over the last two decades, the field of environmental economics has greatly expanded, and a variety of methodologies designed to measure the social values associated with environmental services have been developed.

Weaknesses and Alternative Approaches

Grand Canyon management can be intuitively reduced to a set of decisions about how the Canyon's resources are to be allocated or reallocated. As mentioned above, gains and losses for any allocation decision must be compared. Weighing gains against losses across different groups of people requires that some common metric be chosen so that the units are comparable. Because losses (costs) are counted in dollars, it is common to try to convert gains (benefits) into dollar terms as well. It is not necessary to choose money as the metric, but because costs are usually in dollars, dollars are typically selected as the measure.

There is a natural tendency for many to want to avoid explicitly converting the gains into dollars, especially when environmental goods are involved. Unless this is done, however, explicitly or implicitly, the necessary weighing of gains and losses will involve comparing "apples and oranges." Decision-making is paralyzed until some such comparison is explicitly or implicitly made. At the point where some resource reallocation decision is finally made, it can be inferred that somebody has undertaken to make the conversion, even if only implicitly. It is generally preferable to force transparency upon the decision process by insisting that participants make explicit their assessments of benefits as well as costs.

In many decision-making contexts, including the present one, formal analysis seems to end with an inventory of probable effects of some proposed (or recent) change measured in different physical terms (e.g., a decrease of 10 percent in the population of humpback chub, an increase of 15 percent in the population of rainbow trout, and an increase of 3 percent in average annual electricity prices). It is then left to the ultimate decision-maker to infer which of these physical effects is a gain and which is a loss, who the winners and losers are, and by how much each winner or loser values these effects. These social benefit and cost calculations are typically done informally, without the support of sufficient quantitative research and in sharp contrast to the rigor with which many of the precipitating physical effects are measured.

What needs to be done, and what are the prospects for doing it correctly? For stakeholders who actually use the Grand Canyon, the environmental valuation methods most relevant to Canyon management fall into categories that can be summarized as "travel cost methods" and "contingent valuation." Travel cost methods have a longer tradition. They can still, of course, be implemented badly. They do, however, rely on actual choices made by individuals from which one can infer their willingness to pay through observations of the costs individuals are willing to incur to gain access to the environmental goods in question. Contingent valuation or contingent behavior methods have been far more controversial and are suspected to be more subject to biases because of poor implementation. Information from these methods can be combined with travel cost method information, however, to provide a fuller picture of the choices stakeholders would be likely to make under a variety of both actual and proposed Grand Canyon conditions.

But contingent methods are sometimes the only valuation method that can be used, as in the case of attempting to value changes in the ecological services of a resource where individual values are not "use" values, but "nonuse" values. The overall social values of the ecological services of the Grand Canyon would probably have to be measured in this way by policy makers attempting to compare alternative resource reallocations. Environmental economists distinguish between existence, bequest, or option values for the preservation or enhancement of ecological functions associated with unique natural resources such as the Grand Canyon. These are types of nonuse or passive-use values. That these values are probably positive and substantial is implied by the Grand Canyon's designated status as a World Heritage Site. When nobody is observed to be incurring costs in order to "use" these valuable ecological functions, however, the only recourse is to elicit from individuals, via a general population survey, information about how much they would be willing to pay if a market did exist. Hypothetical valuation exercises are fraught with an inventory of potential biases. Nevertheless, the literature on nonmarket valuation research has been growing in response to the need for estimates of nonmarket, nonuse values in so many contexts. Wetlands policies are an example where the valuation of ecological services has been an important issue (cf. Heimlich et al., 1998).

What are the realistic prospects for measuring everything that needs to be known for a thorough benefit–cost analysis of Grand Canyon management decisions? It would be prohibitively expensive to measure accurately every social benefit and cost associated with some particular suite of physical changes in the Grand Canyon. But it is certainly important that stakeholders be informed about and account for the value imputations they have selected when making recommendations to the Secretary of the Interior on dam-operation alternatives. Implicitly assigned valuations deserve as much scrutiny as the scientifically measured physical effects. Even the best physical measurements can lead to bad management decisions if the social values of these changes are assigned incorrectly. The Strategic Plan contains little discussion of how the Center plans to stay abreast of research on the valuation of nonmarket environmental goods, including both use and nonuse values. More importantly, there is little discussion of how the Center plans to use these valuation methods to monitor the social effects of dam operations.

In some nonmarket valuation contexts, a strategy called "benefits transfer" is highly desirable when feasible. This is a technique of finding other studies done on the values of similar environmental goods, under sufficiently similar conditions, to allow the approximate social values from these other studies to be transferred for use in the current context. Benefits transfer is not likely to be as useful in valuing Grand Canyon resources as it is, say, for valuing the reduction in social value from small oil spills. There have been many small oil spills; there is only one Grand Canyon. For unique resources like the Grand Canyon, benefits transfer is likely to be less fruitful.

In the absence of viable benefits transfer opportunities, it is important to consider the implications of limited budgets for future economic analysis. In the near term, the Center is unlikely to have the internal resources to undertake innovative original survey research to establish social values for different components of the Grand Canyon ecosystem. If future recommendations to the Secretary of the Interior require more precise knowledge about social estimates of environmental benefits than has been needed in the past, it may become necessary to raise funding for research to learn about these benefits. In-house expertise in the relevant environmental valuation methods is a prerequisite for ensuring that the necessary research is done correctly.

Information Technology Program

Figure 4.1 shows a simple model of the flow of data and information, and its role in decision-making in the context of adaptive

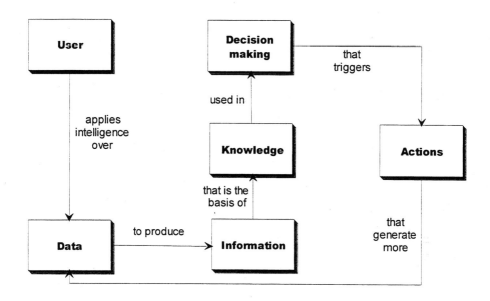

FIGURE 4.1 A model of the data–information–decision-making cycle for the Grand Canyon (adapted from Rob and Coronel, 1997).

management of the Grand Canyon ecosystem. In this model, the Adaptive Management Work Group proposes actions. When the Secretary of the Interior takes an action, the system is monitored: data describing the physical, biological, cultural, and socioeconomic system are collected. These data, when compared with "without action" data, produce information about changes to the ecosystem. This, in turn, provides a basis for judging the efficacy of the action taken, thus leading to further decision-making.

In this model, two "data sets" are equally important: the set that describes the system with the proposed actions taken, and the set that describes the system prior to the action. The Center's scientific programs are charged with monitoring the former and describing the latter, where it is not already done. The Center's Information Technology Program is charged with maintaining and distributing information about the latter (which is, in fact, dynamic because of natural changes in the system).

The Information Technology Program is viewed properly as a *support* program at the Center rather than as a *research* or *monitoring* program. According to the fiscal year 2000 plan, this program's goal is "to satisfy the information needs of stakeholders, scientists, and the public relative to the Colorado River ecosystem." To fulfill this goal, three tasks are assigned to the Information Technology Program:

1. Archiving and delivering scientific data and other information to stakeholders, scientists, and the public.
2. Providing technology-based solutions to data collection, manipulation, and analysis.
3. Providing support in areas of computers, surveying, and geographic information systems.

Task 1: Archiving and delivering scientific data and other information to stakeholders, scientists, and the public.

According to the Center, the Information Technology Program (ITP) "becomes involved with scientific investigations at the point of contract award, to provide relevant background literature, scientific and remotely sensed data, and survey and other spatial data. The researcher identifies to the ITP the type and attributes of...data they are collecting...When GCMRC [Grand Canyon Monitoring and Research Center] receives a deliverable from a researcher...the ITP reviews it...and incorporates it into the appropriate data system [from which it is] made available to stakeholders, researchers, and the public through delivery systems" (Center, 1998).

The Information Technology Program relies on three core technologies for data archiving and delivery:

1. A database management system. A database is a shared, integrated computer structure in which raw facts (data) are filed, along with a description of the characteristics and relationships of the data (metadata). A database management system is a set of software programs that permit a user to manage the database structure, to file and selectively to retrieve data, and to control access to the data.

The Center staff recognizes the value of data and the value of managing these data in a systematic fashion with modern database management systems. They have selected the Oracle database management system as the tool for data management. Implementing this as an enterprisewide system will facilitate: (1) interpretation and presentation of the scientific data in useful formats, (2) distribution of data and information, (3) data preservation and use monitoring, and (4) control over data duplication, internally and externally. Current efforts include installing software, documenting installation, and designing and programming the database structure. Plans for fiscal year 1999 focus on inventorying available data and designing a system for filing these in a consistent electronic format within the Oracle system (implementation of the database management system was scheduled for December 1999 but was delayed because of staff turnover).

2. A geographic information system (GIS). A geographic information system is a software system that integrates the capabilities of a database management system with the capabilities of drawing, drafting, mapping, and coordinate geometry packages. This permits the storage, selective retrieval, and manipulation of data that are spatially referenced, and presentation of the result of the retrieval and manipulation as maps.

Glen Canyon Environmental Studies staff, and subsequently Center staff, recognized the value of the systematic archiving of spatial data and have undertaken work to provide staff, researchers, and stakeholders with GIS capabilities. The Information Technology Program has selected protocols for geographic data storage, and plans for fiscal year 2000 include developing tools for distributing the geographic information system on the Internet, integrating the geographic information system with the database management system, and incorporating data collected in fiscal year 1999.

3. A library. The Center's library is a conventional facility in which books, reports, maps, photographs, and videos are stored and from which these materials are loaned to staff, scientists, and stakeholders. The Information Technology Program manages the library and is responsible for the acquisition and distribution of its holdings. Work is underway to establish policies for library material use and check-out; to catalog contents; to facilitate day-to-day operation; to provide electronic searching capabilities; and to provide more information electronically.

Task 2: Providing technology-based solutions to data collection, manipulation, and analysis.

Scientific data collection, manipulation, and analyses required for Grand Canyon research and monitoring are, in many cases, accomplished best using modern technology. The Information Technology Program is charged with promoting in-house use of this technology. It is also charged with providing coaching and encouragement to stakeholders, outside scientists, and the public in effective use of the technology.

Information Technology Program staff have devoted significant efforts to investigation of remote sensing solutions to the data collection problems, as these solutions promise to provide a cost-effective means of resource monitoring, with minimum impact. The program proposes to allocate approximately 50 percent of its fiscal year 2000 budget to this remote sensing work. Activities will include: (1) evaluation of the utility of satellite and airborne imagery, global positioning systems, telemetry, hydroacoustics, and sonar, (2) acquisition of image-processing software, hardware, and consulting services necessary to make best use of the remotely sensed data, and (3) establishment of ground control for the remotely sensed data (through allocations for topographic and hydrographic surveys).

Other efforts at providing technology-based solutions are intertwined with the database management system and GIS activities that support archiving and delivering scientific data. For example, plans for GIS activities include developing an Internet map server. This relatively new technology will significantly improve the capability of the Center to distribute spatial data to stakeholders so that they can use the information for decision-making.

Task 3: Providing support in areas of computers, surveying, and geographic information systems.

The Information Technology Program supports office automation at the Center. This is a housekeeping task presumably assigned to the Information Technology Program, rather than to administrative staff, because of expertise of the Center's staff with the technology. The Center's system includes approximately 50 computers with various peripherals. The computers are linked within the Center via a local area network and to the world via the Internet.

In addition to this administrative chore, the program provides survey support to researchers. This support includes establishing the location of physical, biological, and cultural features of the Grand Canyon, using global positioning systems, conventional topographic surveying tools, and hydrographic surveys. Products of the survey department include spatial data, which form the basis for various GIS coverage areas, and maps of features of interest. These products are produced for both staff and contractors.

The fiscal year 2000 plan identifies development of protocols for data collection, processing, and use as "areas of focus" for the Information Technology Program. This is critical, for data standards and protocols will ensure consistency in application of technology within the Center and by its contractors. This program has adopted the principles of the National Information Infrastructure, the National Biological Information Infrastructure, and the National Spatial Data Infrastructure, and it has promised to incorporate their guidelines and protocols into the overall database design and into delivery systems whenever possible. This is an important and positive contribution to data maintenance at the Center. As currently programmed, data standard and protocol develop-ment will continue through fiscal year 2000. Other support activities include efforts to provide stakeholders with direct access to selected data and information in the database management system and the GIS, and to assist stakeholders in utilizing data and models incorporated in the Information Technology Program.

Strengths

The roles of the Information Technology Program within the Center are appropriate: the program has not driven the science; it is designed to support it. Its activities are managed much like a business, with goals that can be clearly defined and with performance indicators that can be measured easier and sooner than indicators in the scientific program. The efforts of the Information Technology Program managers to coordinate site surveying in the Grand Canyon have been commendable. Without this, establishing the required geographic references could be chaotic.

Weaknesses and Alternative Approaches

This committee feels that with some modifications, this program could better serve the needs of the stakeholders, scientists, and public relative to the Colorado River ecosystem. These modifications include the following:

1. Survey information users to determine information needs. The stated goal of the Information Technology Program is to "satisfy the information needs of stakeholders, scientists, and the public relative to the Colorado River ecosystem" (Center, 1998). These needs, however, have not been well defined. We thus feel that program staff can contribute significantly to the Center's progress by surveying information users, particularly stakeholders, to identify types of information necessary for informed decision-making and the form in which that information would best be presented. This survey may provide an additional benefit of helping better formulate the questions that are to be answered by the scientific research and monitoring programs.

2. Assign a higher priority to data archiving. Since the earliest reviews of Grand Canyon scientific programs, the lack of archiving of data and results has been criticized. For example, in 1996, the National Research Council committee reviewing the Glen Canyon Environmental Studies wrote that, "Good work was performed and excellent data were collected, but there was little coordination among the different elements of the research team...each project remained essentially an independent entity. There was little coordination of results and little exchange of information among research teams" (NRC, 1996a, p. 74).

This lack of coordination is a communication problem that technology cannot solve. But using technology to archive and distribute data and research results will make coordination easier. For example, if one is interested in studying the movement of cobbles in the river, one should be able to access measurements previously taken without some special "inside track" to locate these data. Researchers at Glen Canyon Environmental Studies reported that they worked with the U.S. Bureau of Reclamation to create metadata reports of all data collected. An electronic metadata form was distributed to all researchers. The goals were to document the data available and to provide a georeference through the Glen Canyon Environmental Studies GIS. To the extent that these metadata reports exist, however, they are not widely available. In fact, the fiscal year 2000 plan notes that "extensive data and information currently exists in the GCMRC...potentially equal amounts...exist within museums, universities, state and Federal agencies, etc. However, much of this information has not been evaluated to assess the interrelationship of resource attributes and differing flow regimes" (Center, 1998).

Various plans lay out programs for information management tasks that may remedy the problem. For example, the 1998 Strategic Plan spells out advantages of using a common database management system. The Oracle system (a good choice) was selected as the enterprise data-warehousing tool, and a plan was developed for implementing the system over several years. But in the meantime, more data will be collected, more scientific research will be conducted, and the volume of data not yet archived will grow.

This committee believes that a carefully formulated strategic plan for database development and management is important. But being correct is of little consequence if the results are too late to influence the decision-making. The delays in database design and implementation put this effort at risk of being too late. The committee thus urges either: (1) adoption of an interim solution that will use available database management tools to make more information available while design and implementation of the enterprise data-warehousing system proceeds, or (2) acceleration of the warehouse development.

We endorse the plan to continue requiring that contributor data be provided in appropriate electronic format. This will expedite data-warehousing and will minimize the risk that newly collected data and results will not be available in a timely fashion to researchers and stake-holders.

According to discussions with this committee, the condition of the Center's library has deteriorated following the transition from Glen Canyon Environmental Studies to the Grand Canyon Monitoring and Research Center. Acquisitions have not been cataloged properly, and loan and recovery of materials have not been monitored carefully. A strategic plan for restoration was developed in October 1998, and a student was employed to assist with this effort. We recommend that this restoration be given higher priority. While much of the academic community is "plugged in" to the Internet and can take advantage of electronic distribution, some stakeholders and large segments of the public cannot. For this group, the documents, photographs, slides, videotapes, and other materials held in the Center's library are critical sources of information.

3. Expand and accelerate data and information delivery via the World Wide Web. The Information Technology Program staff have articulated well the problem that they face: "Bring together years of disparate historical data collected by multiple entities located in databases across the southwest in an organized fashion and then deliver it transparently to an equally disparate group of stakeholders for decision making and modeling purposes" (Center, 1998, p. 77–78).

The Internet, specifically the World Wide Web, provides a partial solution to this problem. Center staff and the U.S. Geological Survey and the Bureau of Reclamation realize this. The main Center Web site (http://www.gcmrc.gov) currently provides information about activities of the Adaptive Management Work Group, the Technical Work Group, and the Center. It permits visitors to download various documents. For example, minutes of the meetings of the Adaptive Management Work Group and Technical Work Group commonly are available. The site also provides access to the annual and long-term monitoring and research plans. Furthermore, the conceptual model (described elsewhere in this report) and accompanying documentation are available for downloading through this site. Recent efforts have presented data (at least a graphical representation of the data), through graphics and animation, of Lake Powell conductivity (see http://www.usbr.gov/gces/pleth.htm on the World Wide Web). Links between the Center and the U.S. Bureau of Reclamation Adaptive Management Program Web pages could be more clearly and closely organized.

The Information Technology Program staff have proposed plans for broader World Wide Web distribution of data from the data warehouse and from the geographic information system, an effort this committee applauds. We feel that much could be done, however, while planning continues. Some relatively quick and inexpensive measures would permit the Information Technology Program to make strides toward satisfying the information needs of stakeholders, scientists, and the public. An example of such an interim solution is the Lake Tahoe data clearinghouse Web site (http://blt.wr.usgs.gov/tahoe/GIS.html#other). This site provides links to databases of several participating federal, state, and local agencies, universities, and tribes. From these sources, a user can retrieve, for example, geographic information system data. In some cases, the link is to a file transfer protocol (FTP) server, such as that at http://edcwww.cr.-usgs.gov/doc/edchome/ndcdb/ndcdb.html. No sophisticated Web interface exists there, and the querying features are limited to "click here if this is what you need." Current data can nevertheless be retrieved in common GIS formats, and with these researchers and stakeholders have access to the information critical for decision-making.

4. Anticipate and plan for development of a computerized decision support system. As described elsewhere in this report, work underway at the Center will contribute to further development of the Grand Canyon ecosystem model. When complete, this conceptual model will provide stakeholders, scientists, and the public with an important opportunity: when used in the context of decision support systems, this model will provide important information for the Adaptive Management Program.

We believe that the Center's Information Technology Program can play a significant role in ensuring that the conceptual model will be a useful tool for scientific investigation, and in promoting the use of the model as a decision support system within the larger Adaptive Management Program. To do this, priorities in the Information Technology Program must be revised to permit staff to interact with the model developers, and to participate in the design and programming to establish data connectivity with the Oracle database management system and the Center's GIS. Current priorities do not permit this. As of early 1999, the database administrator-developer position was vacant and had been vacant for several months. Meanwhile, development of the conceptual model was proceeding quickly, with a projected completion date of March 31, 1999. Opportunities for early coordination of modelers and database developers were thus lost.

Fortunately, the developers of the conceptual model used Microsoft Visual Basic 5.0 as the development tool. Thus, subsequent modifications to the conceptual model by the developers, by the Information Technology Program staff, or by others would be relatively straightforward. Oracle Corporation provides *Oracle Objects for OLE*, a development tool that delivers Oracle database access from Visual Basic, using OLE2 technology. Microsoft provides similar access through *ActiveX Data Objects*. Environmental Systems Research Institute, Inc., provides similar tools for filing, retrieving, and displaying geographic data with Visual Basic applications. With sufficient resources, these applications can be used to provide the conceptual model with access to the Center's databases as the source of state information.

5. Manage computer-system administration independently of the other Information Technology Program activities. The Information Tech-

nology Program staff has recognized that proper system configuration, maintenance, and repair comes at a high cost and has recommended that sources outside the Center administer much of this work. This is possible because such system administration requires no familiarity with specifics of scientific programs. This shift of responsibilities from the staff to a vendor, or to system administrators in the U.S. Geological Survey or the Bureau of Reclamation, will free staff for other duties. In turn, they can concentrate on more important activities that demand familiarity with the Grand Canyon scientific programs.

5

Organization and Resources

Adaptive management programs in the United States are being implemented under a variety of organizational structures, funding arrangements, and resource management settings. Some lessons for successful implementation have been identified (Gunderson et al., 1995). One is that institutional arrangements themselves need to be adaptive, as most attempts to institutionalize adaptive management into a standard template have failed (S. Light, Minnesota Department of Natural Resources, personal communication, 1999). Each setting in which adaptive management is implemented and practiced—its ecosystems, stakeholders, and issues—is complex and unique. Over 40 years ago, Gilbert White cautioned that "No two rivers are the same" (White, 1957, p. 160). Similarly, the structure and organization necessary for success will likely be unique, creating novel structures and procedures over time. At the same time, useful lessons and potential pitfalls may be drawn from past experiences and from analogues with other efforts. A common goal is to maintain and enhance the resiliency of ecosystems and human livelihoods through appropriate management strategies.

We have observed and read about the structure and function of the Adaptive Management Program and have followed the drafting of the Guidance Document. This committee was charged to review whether the Center was functioning effectively in the Adaptive Management Program, which is inextricably linked with other entities in the Program and available resources. This chapter therefore examines the Center's roles in the Program, both as originally envisioned and as

they have evolved. There are many possible organizational arrangements, including the status quo. Nevertheless, as changes may be easier
to effect early in the Program's development, this may be an opportune
moment to recognize potential deficiencies and consider ways in which
they might be resolved. We are sensitive to the significant efforts
invested in the Center and Program, and hope that our recommendations
for improvement are considered in ways that do not negate the
considerable positive efforts to date.

 This chapter begins with a description and assessment of the
Center's roles in the Adaptive Management Program. Recommendations regarding alternatives for the Center's institutional structure,
staffing, and organization are then put forth. We also provide recommendations regarding funding and budget issues in the Adaptive
Management Program that may reduce existing tensions, allowing the
Center and Program to focus more effectively and cooperatively on
ecosystem maintenance and enhancement.

ORGANIZATIONAL STRUCTURE AND CENTER ROLES IN THE ADAPTIVE MANAGEMENT PROGRAM

 The Center is responsible for designing and conducting research
and monitoring activities, ensuring that they meet both the needs of the
Adaptive Management Work Group and the tenets of ecosystem science.
The Adaptive Management Work Group makes recommendations to the
Secretary of the Interior for ecosystem management, based in part upon
the Center's monitoring and research on the effects of Glen Canyon
Dam operating regimes on the ecosystem. These responsibilities were
described in the original Center operating protocols as "consistent and
effective cooperative efforts ongoing in the areas of policy,
administrative and science protocols, definition of research needs, and
dissemination of research information and technology" and as a "close
functional relationship between resource stakeholders and managers and
the Center's science group" (Center, 1996).

 Beyond its monitoring and research programs, the Center has
been expected to be a driving force behind many Adaptive Management
Work Group and Technical Work Group activities. This is contrary to a
model wherein these two groups are responsible for creating a vision of
the Grand Canyon ecosystem and for creating the attendant management

objectives and information needs, with the Center responsible for implementing monitoring and research programs. These activities have been largely defined by the Technical Work Group in coordination with the Center, with final approval resting with the Adaptive Management Work Group.

There thus appears to be a need to revisit the Adaptive Management Program's operational relationships and responsibilities. Without a clarification of roles, it will be difficult for the Center or any entity to document their accomplishments and program rationale in response to the Grand Canyon Protection Act, the Glen Canyon Dam Environmental Impact Statement, and the Record of Decision.

One interesting feature of the Adaptive Management Program has been the establishment of a "management team" within the U.S. Department of the Interior, which regularly discusses a variety of Program issues with Center staff. This team is currently composed of the Deputy Assistant Secretary of the Interior for Water and Science, the Secretary of the Interior's designee in the Adaptive Management Work Group (the current designee has also served as director, Operations, at the U.S. Bureau of Reclamation), the chief hydrologist of the U.S. Geological Survey, and the Center chief. Although this team does not include or represent all stakeholders, and it was instituted to create an ad hoc administrative home for the Center, some consideration should be given to making it permanent, as this would provide a measure of independence and access that supports the intended roles of scientific monitoring and research.

THE CENTER'S INSTITUTIONAL HOME

The Center was temporarily formed under the Office of the Assistant Secretary of the Interior for Water and Science, which provided some autonomy and independence for the monitoring and research programs. This temporary arrangement was recently extended for an additional year. The reality of this arrangement, however, is that there remains a high degree of interdependence between the Center and various agencies. These include payroll and contractual services with the U.S. Bureau of Reclamation and the use of U.S. Geological Survey facilities. The Center, however, should be truly independent if the Program is to conduct truly independent research and monitoring activities.

The Glen Canyon Dam Environmental Impact Statement indicated that the Center would eventually be located in either the U.S. Geological Survey or the U.S. National Biological Survey (since renamed the U.S. Biological Resources Division and integrated into the U.S. Geological Survey).

Several alternatives for the Center's institutional home have been considered. Based on three screening criteria that have been discussed within the Adaptive Management Program, the alternatives that have been considered include the U.S. Bureau of Reclamation, the U.S. Geological Survey, the U.S. National Park Service, as well as extending the current interagency arrangement. Other alternatives that may be considered include a university, an independent science organization such as the Smithsonian Institution, or a new interagency arrangement. All of these alternatives contain a mix of strengths and weaknesses and the committee recognizes the complex and changing situations in each of them. This review and previous National Research Council reports on institutional and administrative issues in the Glen Canyon Environmental Studies indicate that the following criteria, which resemble but extend beyond the screening criteria mentioned above, will be important for making decisions about the Center's institutional home:

1. The Center should be housed within a premier science organization that has a commitment to physical, biological, and social science inquiry.
2. The organization should enable the Center to work effectively with all Grand Canyon and Glen Canyon Dam management agencies.
3. The organization should enable the Center to communicate scientific program issues and results directly with a management team at the Assistant Secretary level in the Department of the Interior.
4. The Center should be independent from any single stakeholder management organization within the Adaptive Management Work Group.

The committee found that no arrangement currently being considered perfectly meets all these criteria. The committee recommends that any proposal for the Center's institutional home within the U.S.

Department of the Interior include an institutional design, addressing institutional constraints and weaknesses related to these criteria.

THE CENTER'S ORGANIZATIONAL STRUCTURE AND SIZE

The Glen Canyon Dam Environmental Impact Statement envisioned the Center as having a small permanent staff of five or six. The Center's initial operations plan from the Office of the Assistant Secretary for Water and Science increased that number to eight to ten permanent staff, with a similar number of temporary positions. Current staff levels are at the upper boundary of that range, with 20–22 positions. The size of the Center's staff and related budget levels have been sources of concern to both the Adaptive Management Work Group and Technical Work Group. Although staff levels have been justified by the Center and approved by the Adaptive Management Work Group, concerns about budget increases remain.

The transition from the Glen Canyon Environmental Studies to the Grand Canyon Monitoring and Research Center involved hiring new staff and keeping some existing staff. Existing staff enhanced the transfer of Glen Canyon Environmental Studies' institutional memory to the Center, while new staff helped initiate needed changes.

The value of having a senior scientist(s) was noted in the 1987 National Research Council committee's report: "no senior scientist or group of experienced science advisors were involved in the early planning or in helping the researchers in analysis and integration during the study. Had experienced scientists been involved, the results almost certainly would have been more satisfactory and useful" (NRC, 1987). A part-time senior scientist was eventually hired (1989–1996) and a draft integration report was prepared (in 1998).

A senior scientist could again help ensure that current efforts fit both the ecosystem science paradigm and applied needs of the Adaptive Management Work Group. The committee recommends this position should be created and filled, as it was previously filled at the Glen Canyon Environmental Studies. Given its roles in both facilitating the Adaptive Management Program and implementing research and monitoring programs, the Center needs a different management structure. Earlier recommendations called for a position of senior scientist to help keep a focus on ecosystem science and research. This recommendation was followed in the past, with the post filled on a part-time basis. Given the broader range of stakeholder involvement in the Program, the pressing need to implement the monitoring program, and the gap in research integration and synthesis, the committee recommends the appointment of a full-time senior scientist. The committee recommends that the senior scientist enjoy a high degree of independence (e.g., reporting directly to the Secretary of the Interior's office). The committee believes this independence would help promote an interest in the adaptive management experiments and help attract the interest of widely recognized scientists in the position. This senior scientist may represent the best means of ensuring synthesis and integration of information in the Adaptive Management Program. The senior scientist would also help articulate adaptive management experiments, including hypotheses, experimental treatments, and expected outcomes.

In addition to promoting an ecosystem perspective and articulating the current adaptive management experiment—which would benefit both scientists and managers—a senior scientist would be well-placed to help develop an ecosystem vision (see Chapter 3) and serve as an effective advocate for the adaptive management experiments themselves. This would help represent the integrity and consistency of the experiments before all parties, scientists, managers, and the public.

The 1987 National Research Council review also suggested that it was unlikely that an administrative director (then of the Glen Canyon Environmental Studies, now of the Center) would be able to simultaneously fulfill the demanding roles of science administrator and science visionary: "There was no clear separation of administrative and scientific oversight for the GCES project...the GCES project manager was also one of the researchers, the contract manager, and the report integrator, and was looked to for general oversight...the committee believes that no one person should have been assigned such diverse responsibilities for research and management in such a large environmental study" (NRC, 1987). This committee finds these conclusions to apply equally as well today.

The Center has sought a balance between its ability to contract research and monitoring activities and to conduct research and monitoring in-house. Maintaining both capabilities is a challenge. Research scientists are most knowledgeable about the Grand Canyon ecosystem, but they typically do not make good contract officers (and vice versa).

The Center and Program are well-served by the current cadre of scientists. In general, each resource program should have at least one staff member with scientific expertise and another with administrative skills.

Additional staff and associated budget allocations seem warranted for the existing Physical Resources, Cultural Resources, and Socioeconomic Resources programs. These programs presently have only one or no staff. For example, socioeconomic analysis warrants an additional person. The Biological Resources Program is currently well staffed. Despite concerns voiced about increases in the number of Center staff, staff expertise is necessary for evaluating policy trade-offs, decision analysis, and adaptive management planning.

A related organizational and staffing issue has emerged because of the twin roles played by the Center in Program planning and scientific research. Although these twin responsibilities were anticipated in the Center's original operating protocols, the primary emphasis was on science-based research and monitoring. The current organizational structure has thrust science researchers and managers into roles of program-wide planning.

The Center's original operating protocols (developed in 1996) stated, "Ecosystem science, although becoming more prominent in government science programming, is still in a developmental stage. Merging of the adaptive management procedure with ecosystem science methodology creates a science planning and implementation paradigm that is even less developed. An important outcome from this program will be improved design and operational procedures for merging adaptive management and ecosystem science concepts" (Center, 1996).

In addition to a senior scientist, there is a need for an adaptive management specialist at the Center. This specialist's roles would include the explicit incorporation of adaptive management planning within the Center and the Program. The adaptive management specialist would have knowledge of institutional aspects of adaptive management and skills in policy analysis. This person would, among other tasks, help identify and articulate links between scientific research, alternatives analysis, and adaptive management processes. The committee feels that both these positions are essential to the successful execution of a science-based, ecosystem-level, adaptive management program associated with Glen Canyon Dam operations and their downstream effects.

BUDGET AND FUNDING ISSUES

The budget for the Adaptive Management Program has been in the general range of $7–7.5 million (Table 5.1). These funds, as provided for in the Grand Canyon Protection Act, come from sales of hydroelectricity through the Western Area Power Administration (WAPA). Table 5.1 indicates the budget for the Center's monitoring and research programs, and the administrative costs of the Bureau of Reclamation and the Programmatic Agreement. The Center has also occasionally sought additional funds for research and monitoring. Some Adaptive Management Work Group and Technical Work Group members have expressed concerns over proposed increases to budget and staff, as well over increases in the Adaptive Management Program's geographic scope. This is understandable, as the revenue for the Program comes directly from activities in which they have a vested interest.

The Grand Canyon Protection Act allows for funding of research and monitoring programs from power revenues; however, it neither requires nor precludes funding from other sources. In fact, Center scientists have obtained outside funding for some projects. There are reasonable questions regarding the funding of all Program activities through a single source. While one could argue that these are federal funds, their use nevertheless affects some stakeholders, and others not at all. It can also be argued that this funding is reasonable, as the Glen Canyon Dam and its operations have caused most of the changes being investigated and monitored.

It may be useful to recall how research and monitoring activities have been classified as "white," "gray," and "black." In the opinion of many Adaptive Management Work Group members, white issues are the only ones that clearly fall under the responsibility of the Adaptive Management Program. As one proceeds to the gray and black issues there is less agreement, not necessarily about the value of the research, but whether it should be funded under the current arrangement. It seems reasonable that a core program of staffing and research be established at current levels or greater and that some long-term assurance be provided regarding the stability of these funds. One can then refine the criteria for determining which additional future activities should be supported from

TABLE 5.1 Grand Canyon Research and Monitoring Center Budget (values in millions of dollars)

	FY 1998	FY 1999	FY 2000*
Adaptive Management Program Administration and Support	1.4	1.4	1.4
Center			
Bureau of Reclamation support			.1
Operations and Personnel	1.9	1.9	2.0
Physical Resources	1.2	1.2	.7
Biological Resources	1.4	1.4	1.5
Cultural Resources	.4	.4	.3
Socioeconomic Resources	.6	.6	.06
Information Technology	.4	.4	.3
Other, including remote sensing technology, logistics, and independent review			1.2
TOTAL	~7.3	~7.3	~7.7

* the fiscal year 2000 budget is one of a few of the proposed budget estimates. While the figures are thus not final, they are indicative of evolving allocations within the Program.

SOURCE: Center (1998).

additional power revenue funds and which might come from the budgets of other agencies (e.g., in the U.S. Department of the Interior and foundations).

This committee feels that core funding at least at the level currently provided is essential, but that there should be both flexibility and encouragement for the Center and its collaborating scientists to seek additional funds. We also believe the Strategic Plan should provide for some form of budget escalation to offset inflation. This will be a long-term program, and the funding commitment should reflect that fact. A multiyear funding arrangement coordinated through the agencies and Congress should be considered. This could also ensure more stability for future monitoring and research needs (see NRC, 1996a).

The committee also believes that performance and fiscal responsibility are important in this program and that costs need to remain reasonable. To fulfill the aims of the Grand Canyon Protection Act and the Secretary of the Interior's related responsibilities, however, it would behoove the Adaptive Management Program to find ways to enhance the program's fiscal resources as needed, and to reduce the impediments created by the current funding arrangements.

6

Summary of Findings and Recommendations

The Grand Canyon Monitoring and Research Center is engaged in a major science-policy experiment in western U.S. water management. It is one of the only comprehensive science organizations designed to support an adaptive management program. The U.S. Department of the Interior's Adaptive Management Program is pioneering in many respects. The Program has given rise to legal changes in Glen Canyon Dam operations, policy and science program decisions for the Colorado River ecosystem are based upon direct stakeholder input, and it is recognized that future Glen Canyon Dam operations may need to be continually adjusted in response to changing scientific knowledge and public values. Changing values in the 1970s and 1980s, and surprising environmental results of floods in the early 1980s, led to the establishment and continuation of the Glen Canyon Environmental Studies. Scientific findings from that program demonstrated that Glen Canyon Dam operations had significant effects on downstream resources. The Grand Canyon Protection Act of 1992, the Glen Canyon Dam Environmental Impact Statement (1995), and the Secretary's Record of Decision (1996) led to the establishment of the Adaptive Management Program, which includes the Grand Canyon Monitoring and Research Center.

This National Research Council committee was convened to assess the Strategic Plan's likely effectiveness in meeting the requirements specified in the above-listed mandates. More specifically, the committee was asked to address two main questions and five related questions regarding the Center's Strategic Plan:

1.   Will the Long-Term Strategic Plan be effective in meeting requirements specified in the Grand Canyon Protection Act, the final Glen Canyon Dam Environmental Impact Statement, and Record of Decision?

a.   Does the Long-Term Plan respond to the new adaptive management process called for by the Grand Canyon Protection Act and Glen Canyon Dam Environmental Impact Statement?   Is the Grand Canyon Monitoring and Research Center functioning effectively in the Adaptive Management Program, especially regarding incorporation of all stakeholder objectives and information needs in the planning process?
b.   Does the Long-Term Plan incorporate past research knowledge in developing new monitoring and research directions?
c.   Has the Center appropriately addressed past reviews of Glen Canyon Environmental Studies programs in formulating new research directions?

2.   Characterize weaknesses of the Long-Term Plan and recommend short and long-term science elements to the GCMRC to address identified weaknesses.

a.   What weaknesses exist in the Long-Term Plan, and how do these weaknesses affect the potential effectiveness of the overall science program?
b.   What science elements are necessary to correct specific plan weaknesses?

In addition to reviewing the Strategic Plan, the committee was asked to comment upon the Center's functions within the larger Adaptive Management Program (as described within the Grand Canyon Protection Act and the Glen Canyon Dam Environmental Impact Statement).

The Center's Strategic Plan has a good chance of fulfilling mandated requirements. Although the requirements of these federal acts and documents are still being clarified, and the Strategic Plan is being revised, the Center has made important strides toward establishing an effective monitoring and research program. The Center has also responded well to the new Adaptive Management Program. The chances of meeting national policy aims and requirements will be enhanced if the following recommendations are addressed.

ADAPTIVE MANAGEMENT ISSUES

• **Begin the long-term monitoring program.** The Center is in a good position to start this program and should implement it in the near future.

• **Clarify the scientific basis for the adaptive management experiment currently being conducted in the Grand Canyon.** The hypothesized relations between dam operations, ecosystem responses, and social effects must be defined.

• **Develop a more sophisticated and flexible definition of the geographic scope of the Adaptive Management Program.** The Program and the Adaptive Management Work Group have found ways to creatively address boundary issues (e.g., Lake Powell). Future boundary issues should also be thoughtfully and flexibly addressed.

• **Include a strategy for scientific evaluation of policy alternatives, both in terms of ecological outcomes and values of stakeholder groups.** The Adaptive Management Program's strategic plan should include a strategy for using new scientific information in drafting policy options.

• **Recognize limitations of the current pluralistic situation within the Adaptive Management Program.** The Center and the Adaptive Management Work Group should work together to identify a set of baseline conditions and vision for the Grand Canyon ecosystem.

• **Continue to work toward a set of internally consistent, refined, and reduced management objectives and information needs.** These should be created through collaboration between the Center, a new senior scientist, and the Adaptive Management Work Group.

• **Explicitly recognize that effective adaptive management in the Grand Canyon will require trade-offs among management objectives favored by different groups. The Adaptive Management Work Group should begin to consider mechanisms for equitable weighting of competing interests. The Center should begin to develop decision support systems and methods.**

• **To ensure credible, objective review of the Center and the Program, establish a Science Advisory Board that is not a subcommittee of the Adaptive Management Work Group. Issues addressed by the Science Advisory Board should not be formally limited.**

SCIENCE PROGRAM ISSUES

In all the Center's science programs, it is important that Center scientists and stakeholders play a role in identifying research needs. The selection and design of appropriate scientific investigations within the Adaptive Management Program should be guided both by competitive requests for proposals and by advice from independent review panels.

- **Core monitoring variables should be explicitly identified and should consist of simple and basic information whose value will accrue over time.** These data should be selected using ecosystem-level, multispecies perspectives. Monitoring programs must be shielded from fluctuating budgets and short-term interests.
- **There is a need for more and better knowledge regarding sediment budgets, particularly in upstream reaches impacted by post-dam supply reductions, and in Glen and Marble canyons.**
- **Biological research should be shifted from its present species-oriented emphasis toward broader monitoring and research on communities and ecosystems. It must also address the biological implications of the temperature-control experiments involving selective withdrawals from Lake Powell.**
- **The Cultural Resources Program should look forward to encompassing a broader range of social groups and historical periods, and to recognizing that tribal perspectives and cultural resources provide valuable insights into adaptive environmental management in the Grand Canyon. Resources for full tribal participation in monitoring, research, and adaptive management must be secured, without reducing other components of the Center's Cultural Resources Program.**
- **The Center should develop expertise and budgeting for modern techniques of nonmarket valuation of ecosystem services.** The scope of economics inquiry in the Strategic Plan is out of balance with the level of research on other features in the Grand Canyon ecosystem.
- **The Strategic Plan and Center should seek to understand not simply the range of preferences and activities of Grand Canyon "users," but also the degree to which the uses and ecosystem features are valued.**
- **Sources of funding for original research devoted to measuring Grand Canyon ecosystem values should be sought, using a fully representative scientific sample of all stakeholders.**

One of the Strategic Plan's strengths is its understanding of theories and practices of adaptive management. In future versions it should anticipate the need to assess the actual uses of results from research and monitoring. The Center's incorporation of past research varies from very good (e.g., physical) to weak (socioeconomic). Regarding previous reviews of the Glen Canyon Environmental Studies, results are similarly mixed. The Center's responses to earlier advice regarding cross-program integration within an ecosystem framework are partially adequate, while responses in the fields of socioeconomic and decision analysis represent backward steps.

ORGANIZATIONAL AND BUDGET ISSUES

- **The operational relationships and responsibilities of organizational entities within the Adaptive Management Program should be reexamined.** There is a current trend toward micromanagement of the Center's activities.
- **The following criteria should be considered in deciding upon the Center's institutional home: (1) the Center should be housed in a premier science organization committed to physical, biological, and social science inquiry, (2) the institutional home should enable the Center to work effectively with all Grand Canyon and Glen Canyon Dam management agencies, (3) the institutional home should enable the Center to communicate scientific program issues and results directly with a management team at the Assistant Secretary level in the Department of the Interior, and (4) the Center should be independent from any single stakeholder management organization within the Adaptive Management Work Group.**
- **A senior scientist and an adaptive management specialist should be appointed to the Center's staff. Additional staff and associated budget allocations also seem warranted for the Physical Resources, Cultural Resources, and Socioeconomic Resources programs.**
- **The Program should consider using hydropower revenues at least at the levels currently provided to support core research, monitoring, and adaptive management programs required by the Grand Canyon Protection Act, the Glen Canyon Dam Environmental Impact Statement, and the Record of Decision. Budgets for additional future activities could be developed from other U.S. Department of the**

Interior agencies and foundation sources, as well as hydropower revenues.

For the Center to be more effective in responding to the Adaptive Management Program and its stakeholder groups, it will need to address recommendations regarding organizational and budget issues.

Future revisions of the Strategic Plan will hopefully focus upon the recommendations listed above and elaborated upon in this report, especially the following: clearly define the adaptive management experiment; implement the monitoring program; conduct monitoring within an ecosystem (vs. species-oriented) paradigm; review the Center's resource programs, responsibilities, and relations with other entities within the Adaptive Management Program; resume socioeconomic analysis and decision support; broaden the definitions of cultural groups and economic resources; and secure broad, objective program review.

The Grand Canyon Monitoring and Research Center and the Adaptive Management Program have made important progress toward management of the Glen Canyon Dam and the Colorado River ecosystem based upon ecosystem science and input from a range of constituencies. The committee commends all involved for their contributions toward these vital trends in water resource management and science-policy innovations, and we look forward to the future ecological and social benefits of strategic planning efforts currently underway at the Grand Canyon Monitoring and Research Center.

References

Adaptive Environmental Assessment Steering Committee and Modeling Team. 1997. Upper Mississippi River Adaptive Environmental Assessment. Phase I report.

American Geophysical Union (AGU). 1999. The Controlled Flood in Grand Canyon. R. H. Webb, J. C. Schmidt, G. R. Marzolf, and R. A. Valdez, eds. Geophysical Monograph Series, Vol. 110. Washington, D.C.: American Geophysical Union.

Anderson, B. W. and R. D. Ohmart. 1985. Managing riparian vegetation and wildlife along the Colorado River: Synthesis of data, predictive models and management. In Riparian Ecosystems and their Management: Reconciling Conflicting Uses, R. R. Johnson, et al., Technical Coordinators, U. S. Foreign Service Technical Report RM-120: 124-127.

Andrews, E. D., C. E. Johnson, J. C. Schmidt, and M. Gonzales. 1999. Topographic evolution of sand bars. In The Controlled Flood in Grand Canyon, R. H. Webb, J. C. Schmidt, G. R. Marzolf, and R. A. Valdez, eds. Geophysical Monograph Series, Vol. 110, Washington, D.C.: American Geophysical Union.

Argonne National Laboratory. 1999. Testing the Effects of a Short-Duration 60,000 cfs Spike Flow and Subsequent Load-Following Operations at Glen Canyon Dam. Report prepared for Western Area Power Administration, Salt Lake City, Utah.

Bennett, J. 1969. Northern Plainsmen: Adaptive Strategy and Agrarian Life. Arlington Heights, Ill.: AMH Publishing Corp.

Berry, J., G. D. Brewer, J. C. Gordon, and D. R. Patton. 1998. Closing the gap between ecosystem management and ecosystem research. Policy Sciences 31:55–80.

Bravo, C., and R. Susanyatame. 1997. Hualapai Tribe Stakeholder Perspective. Pp. 62–63 in Colorado River Basin Management Study. Final report. Flagstaff, Ariz.: Grand Canyon Trust.

Brown, D. E., ed. 1994. Biotic Communities: Southwestern United States and Northwestern Mexico. Salt Lake City: University of Utah Press.

Brown, B. T., S. W. Carothers, and R. R. Johnson. 1987. Grand Canyon Birds: Historical Notes, Natural History, and Ecology. Tucson: University of Arizona Press.

Brunner, R. D., and T. W. Clark. 1997. A practice-based approach to ecosystem management. Conservation Biology 11:48–58.

Butzer, K. W. 1989. Cultural ecology. Pp. 192–208 in Geography in America. G. Gaile and C. Wilmott, eds. Columbus, Ohio: Merrill.

Callan, S. J., and J. M. Thomas. 1996. Environmental Economics and Management: Theory, Policy, and Applications, Chicago, Ill.: Irwin.

Carothers, S. W., and S. W. Aitchison. 1976. An Ecological Survey of the Riparian Zone of the Colorado River Between Lees Ferry and the Grand Wash Cliffs, Arizona. Technical Report No. 17. Grand Canyon, Ariz.: U.S. National Park Service, Grand Canyon National Park.

Carothers, S. W., and C. O. Minckley. 1981. A survey of the aquatic flora and fauna of the Grand Canyon. Final Report. U.S. Water and Power Resources Service. Boulder City, Nev.: Flagstaff, Ariz.: Museum of Northern Arizona.

Carothers, S. W., and B. T. Brown. 1991. The Colorado River Through Grand Canyon: Natural History and Human Change. Tucson: University of Arizona Press.

Carothers, S. W., and D. A. House. 1996. The role of science in Colorado River management. Pp. 155–174 in The Colorado River Workshop: Issues, Ideas, and Directions. Phoenix, Ariz.: Grand Canyon Trust.

Center (GCMRC). 1996. Operating Protocols for Grand Canyon Monitoring and Research Center. U.S. Department of the Interior, Office of Assistant Secretary for Water and Science. Flagstaff, Ariz.: Grand Canyon Monitoring and Research Center.

Center (GCMRC). 1997. The Grand Canyon Monitoring and Research Center Long-Term Monitoring and Research Strategic Plan. Flagstaff, Ariz.: Grand Canyon Monitoring and Research Center.

Center (GCMRC). 1997a. Glen Canyon Dam Beach/Habitat-Building Flow Symposium. Abstracts and Executive Summaries. Flagstaff, Ariz.: Grand Canyon Monitoring and Research Center.

Center (GCMRC). 1998. Second draft of the revised Long–Term Strategic Plan (November 1998). Flagstaff, Ariz.: Grand Canyon Monitoring and Research Center.

Center (GCMRC). 1999. Colorado River Ecosystem Science Symposium 1999. Abstracts, February 16–17, at Grand Canyon National Park. Flagstaff, Ariz.: Grand Canyon Monitoring and Research Center.

Clarkson, R. W., O. T. Gorman, D. M. Kubly, P. C. Marsh, and R. A. Valdez. 1994. Management of Discharge, Temperature, and Sediment in Grand Canyon for Native Fishes. Flagstaff, Ariz. (privately published).

Comprensive Monitoring Assessment and Research Program (CMARP) Steering Committee. 1998. A proposal for the development of a comprehensive monitoring and assessment. Proposal developed for CALFED. Sacramento, Calif.

Denevan, W. M. 1983. Adaptation, variability and cultural geography. The Professional Geographer 35(4):399–406.

Deputy Assistant Secretary for Water and Science. 1995. Establishment of the Grand Canyon Monitoring and Research Center. Washington, D.C.: U.S. Department of the Interior.

Dewey, J. 1938. Logic: The Theory of Inquiry. New York: Holt, Rinehart, Winston.

Dewey, J. 1958. Experience and Nature. New York: Dover Publications, Inc.

Dongoske, K. 1996. Integrating Native American economic and cultural interests into the Colorado River basin. In The Colorado River Workshop. Phoenix, Ariz.: The Grand Canyon Trust.

Dongoske, K., and M. Yeatts. 1998. The Integration of Glen Canyon Dam Programmatic Agreement with the Adaptive Management Program: A Discussion Paper. Final version November 11, 1998.

Douglas, M. E., and P. C. Marsh. 1996. Population estimates/population movements of Gila cypha, an endangered cyprinid fish in the Grand Canyon region of Arizona. Copeia 1996:15–28.

Douglas, M. E., and P. C. Marsh. 1998. Population and survival estimates of Catostomus latipinnis in northern Grand Canyon, with distribution and abundance of hybrids with Xyrauchen texanus. Copeia 1998:915–925.

Ellen, R. 1982. Adaptation: A Summary and Reconsideration. Pp. 236–51 in Environment, Subsistence and System: The Ecology of Small Scale Social Formations. Cambridge: Cambridge University Press.

Fairley, H. C. et al. 1994. The Grand Canyon River corridor survey project: archaeological survey along the Colorado River between Glen

Canyon dam and Separation Canyon. Cooperative Agreement No. 9AA-40-07920. Flagstaff, Ariz.: Glen Canyon Environmental Studies.

Forman, R. T. 1996. Land Mosaics: The Ecology of Landscapes and Regions. Cambridge: Cambridge University Press.

Ferguson, T. J. 1998. Ongtupqa niqw Pisisvayu (Salt Canyon and the Colorado River). The Hopi People and the Grand Canyon. Produced by the Hopi Cultural Preservation Office, under the guidance of the Hopi Cultural Resources Advisory Task Team, and under contract with the U.S. Bureau of Reclamation. Tucson, Ariz.: Anthropological Research.

Frederickson, L. 1996. Exploring Spiritual Benefits of Person-Nature Interactions Through an Ecosystem Management Approach. Ph.D. dissertation. Minneapolis: University of Minnesota.

Graf, W. 1985. The Colorado River: Instability and Basin Management. Washington, D.C.: Association of American Geographers.

Grams, P. E., and J. C. Schmidt. 1999. Integration of Photographic and Topographic Data to Develop Temporally and Spatially Rich Records of Sand Bar Change in the Point Hansbrough and Little Colorado River Confluence Study Reaches. Flagstaff, Ariz.: Grand Canyon Monitoring and Research Center.

Griffiths, P. E. 1996. The historical turn in the study of adaptation. British Journal of Philosophy of Science 47:511-32.

Gunderson, L., C. S. Holling, and S. S. Light, eds. 1995. Barriers and Bridges to the Renewal of Ecosystems and Institutions. New York: Columbia University Press.

Halbert, C. L. 1993. How adaptive is adaptive management? Implementing adaptive management in Washington State and British Columbia. Reviews in Fish Biology and Fisheries 1:261–83.

Haney, A., and R. Power. 1996. Adaptive management for sound ecosystem management. Environmental Management 20: 879–886.

Hanley, N., J. F. Shogren, and B. White. 1997. Environmental Economics in Theory and Practice. New York: Oxford University Press.

Harris, C. S. 1998. Overview of the law of the Colorado River: A historical perspective of the legal and physical operations of the Colorado River. Symposium and Workshop on Restoring Natural Function within a Modified Riverine Environment, Las Vegas, Nev. July 8–9, 1998.

Hart, E. R. 1995. Zuni and the Grand Canyon: A Glen Canyon Environmental Studies Report. Zuni Glen Canyon Environmental Studies Ethnohistorical Report. Seattle, Wash.: Institute of the North American West.

Hastings, J. R., and R. M. Turner. 1964. The Changing Mile: An Ecological Study of Vegetation Change with Time in the Lower Mile of an Arid and Semiarid Region. Tucson: University of Arizona Press.

Harwell, M. A. 1998. Science and environmental decision-making in South Florida. Ecological Applications 8(3):580–590.

Hazel J. E., Jr., M. Kaplinski, R. Parnell, M. Manone, and A. Dale. 1999. Topographic and bathymetric changes at thirty-three long-term study sites. In The Controlled Flood in Grand Canyon, R. H. Webb, J. C. Schmidt, G. R. Marzolf, and R. A. Valdez, eds. Geophysical Monograph Series, Vol. 110. Washington, D. C.: American Geophysical Union.

Healey, M., W. Kimmerer, G. M. Kondolf, R. Meade, P. B. Moyle, and R. Twis. Strategic Plan Core Team for the CALFED Bay-Delta Program. 1998. Strategic Plan for the Ecosystem Restoration Program. Prepared for the CALFED-Bay Delta Program. Sacramento, Calif.

Heimlich, R. E., et al. 1998. Wetlands and agriculture: Private Interests and Public Benefits. Economic Research Service Report No. 765. Washington, D.C.: U.S. Department of Agriculture.

Hoffmeister, D. F., 1971. Mammals of Grand Canyon. Urbana: University of Illinois Press.

Holling, C. S., ed. 1978. Adaptive Environmental Assessment and Management. New York: John Wiley and Sons.

Holling, C. S. 1998. Novelty, rigor and diversity. Conservation Ecology. 2(2). [http://www.consecol.org/journal/vol2/iss2/art14].

Houck, O. A. 1998. Are humans part of ecosystems? Environmental Law 28:1–14.

Howard, A., and R. Dolan. 1981. Geomorphology of the Colorado River in the Grand Canyon. Journal of Geology 89 (3):269–298.

Independent Scientific Group. 1996. Return to the River. Restoration of Salmonid fishes in the Columbia River ecosystem. Development of an Alternative Conceptual Foundation and Review and Synthesis of Science underlying the Columbia River Basin Fish and Wildlife Program of the Northwest Power Planning Council. Prepublication copy. NWPPC Report 96-6. Boise, Idaho: Northwest Power Planning Council.

Ingram, H., A. D. Tarlock, and C. R. Oggins. 1991. The law and politics of the operation of Glen Canyon Dam. In Colorado River Ecology and Dam Management. Proceedings of a Symposium, May 23–25, 1990, Santa Fe, N. Mex. Washington, D.C.: National Academy Press.

International Union for the Conservation of Nature and The World Bank Group. 1997. Large Dams: Learning from the Past, Looking at the

Future. Gland, Switzerland and Washington, D.C.: IUCN and The World Bank.

Jassby, A., C. Goldman, J. Reuter, and R. Richards. 1999. Origins and scale-dependence of temporal variability in the transparency of Lake Tahoe, California-Nevada. Limnology and Oceanography 44:282-294.

Johnson, R. R. 1977. Synthesis and management implications of the Colorado River Research Program. Technical Report No. 10. Grand Canyon, Ariz.: U.S. National Park Service, Grand Canyon National Park.

Johnson, R. R. 1991. Historic changes in vegetation along the Colorado River in the Grand Canyon. In Colorado River Ecology and Dam Management. Proceedings of a Symposium, May 23–25, 1990, Santa Fe, N. Mex. Washington, D.C.: National Academy Press.

Kearsley L. H., et al. 1999. Changes in the number and size of campsites as determined by inventories and measurements. In The Controlled Flood in Grand Canyon, R. H. Webb, J. C. Schmidt, G. R. Marzolf, and R. A. Valdez, eds. Geophysical Monograph Series, Vol. 110. Washington, D.C.: American Geophysical Union.

Keiter, R. B. 1994. Beyond the boundary line: Constructing a law of ecosystem management. University of Colorado Law Review 65:293ff.

Kelley, K. B., and H. Francis. 1994. Navajo Sacred Places. Bloomington: Indiana University Press.

Knaap, G., and T. Kim. 1998. Environmental Program Evaluation: A Primer. Urbana and Chicago: University of Illinois Press.

Korman, J., and C. Walters. 1998. User's Guide to the Grand Canyon Ecosystem Model. Flagstaff, Ariz.: Grand Canyon Monitoring and Research Center.

Lavender, D. 1985. River Runners of the Grand Canyon. Grand Canyon, Ariz.: Grand Canyon Natural History Association.

Lee, K. 1993. Compass and Gyroscope: Integrating Science and Politics for the Environment. Covelo, Calif.: Island Press.

Lewis, W. M., Jr. 1994. The ecological sciences and the public domain. University of Colorado Law Review 65:279–292.

Light, A., and E. Katz. 1996. Environmental Pragmatism. London: Routledge.

Lindblom, C. 1959. The science of muddling through. Public Administration Review 19: 79–88.

Ludwig, D., R. Hilborn, and C. J. Walters. 1993. Uncertainty, resource exploitation, and conservation: Lessons from history. Science 260:17–

36.

Machlis, G. E., J. E. Force, and W. R. Burch, Jr. 1997. The human ecosystem, Part I: The human ecosystem as a organizing concept in ecosystem management. Society and Natural Resources 10:347–367.

Marsh, P. C., and M. E. Douglas. 1997. Predation on endangered humpback chub and other native species by introduced fishes in the Little Colorado River, Arizona. Transaction of the American Fisheries Society 126(2):343–346.

Martin, R. 1990. A Story that Stands Like a Dam: Glen Canyon and the Struggle for the Soul of the West. New York: Henry Holt.

Marzolf, G. R. 1991. The role of science in natural resource management: The case for the Colorado River. In Colorado River Ecology and Dam Management. Proceedings of a Symposium, May 23-25, 1990, Santa Fe, N. Mex. Washington, D.C.: National Academy Press.

Marzolf, R. G., R. A. Valdez, J. C. Schmidt, and R. H. Webb. 1998. Perspectives on river restoration in the Grand Canyon. Bulletin of the Ecological Society of America 79(4): 250–254.

Mastrop, H. and A. Faludi. 1997. Evaluation of strategic plans: the performance principle. Environment and Planning B: planning and design 24: 815–832.

May, P. J. et al. 1996. Environmental Management and Governance: Intergovernmental Approaches To Hazards and Sustainability. London: Routledge.

Melis, T. 1998. Draft Grand Canyon Monitoring and Research Center response to the Glen Canyon Technical Work Group (ad hoc group). Request for assessment of A Proposal to Develop a Research Plan to Analyze Resource Responses to Alternative BHBF and Load-Following Releases from Glen Canyon Dam. Flagstaff, Ariz.: Grand Canyon Monitoring and Research Center.

Minckley, W. L. 1979. Aquatic habitats and fishes of the lower Colorado River, Davis Dam to the U.S.–Mexican International Border. Final Report. Boulder City, Nev.: U.S. Bureau of Reclamation.

Minckley, W. L. 1985. Aquatic habitats and fishes of U.S. Fish and Wildlife Service Region II west of the Continental Divide. Final Report. Albuquerque, N. Mex.: U.S. Fish and Wildlife Service.

Minckley, W. L. 1991. Native fishes of the Grand Canyon region: An obituary? In Colorado River Ecology and Dam Management. Proceedings of a Symposium, May 23–25, 1990, Santa Fe, N. Mex. Washington, D.C.: National Academy Press.

Mintzberg, H. 1990. Strategy formation: Schools of thought. In Perspectives on Strategic Management. J. W. Frederickson, ed. New

York: Harper Business.

Mintzberg, H., B. Ahlstrand, and J. Lampel. 1998. Strategy safari: a guided tour through the wilds of strategic management. New York: The Free Press.

Mintzberg, H., and J. Waters. 1998. Of Strategies, Deliberate and Emergent. In The Strategy Reader. S. Segal-Horn, ed. London: Blackwell.

Moote, M. A., S. Burke, H. J. Cortner, and M. G. Wallace. 1994. Principles of Ecosystem Management. Tucson: University of Arizona Water Resources Research Center.

Morehouse, B. J. 1996. A Place Called Grand Canyon: Contested Geographies. Tucson: University of Arizona Press.

National Academy of Sciences. 1968. Water and Choice in the Colorado Basin: An Example of Alternatives in Water Management. Washington, D.C.: National Academy of Sciences.

National Research Council. 1987. River and Dam Management: A Review of the Bureau of Reclamation's Glen Canyon Environmental Studies. Washington, D.C.: National Academy Press.

National Research Council. 1991. Colorado River Ecology and Dam Management. Proceedings of a Symposium, May 24–25, 1990, Santa Fe, N. Mex. Washington, D.C.: National Academy Press.

National Research Council. 1992. Long-Term Monitoring Workshop for the Grand Canyon. Irvine, Calif.

National Research Council. 1994. The Glen Canyon Environmental Studies. Review of the Draft Federal Long-Term Monitoring Plan for the Colorado River Below Glen Canyon Dam. Washington, D.C.: National Academy Press.

National Research Council. 1996a. River Resource Management in the Grand Canyon. Washington, D.C.: National Academy Press.

National Research Council. 1996b. Upstream: Salmon and Society in the Pacific Northwest. Washington, D.C.: National Academy Press.

National Research Council. 1998. Cooperative Agreement No.1425-98-FC-40-22700 for Review of the Adaptive Management Work Group Strategic and Annual Plans. NRC Water Science and Technology Board and U. S. Department of the Interior.

Nyberg, J. B. 1998. Statistics and the practice of adaptive management. Pp. 1-9 in Statistical Methods for Adaptive Management Studies. V. Sit and B. Taylor, eds. Land Management Handbook 42. Victoria, British Columbia: Ministry of Forests.

Ohmart, R. D., B. W. Anderson, and W. C. Hunter. 1988. The ecology of the lower Colorado River from Davis Dam to the Mexico–United

States International Boundary: A community profile. Tempe: Arizona State University, Center for Environmental Studies.

O'Neill, R. V., D. L. DeAngelis, J. B. Waide, and T. Allen. 1986. A Hierarchical Concept of Ecosystems. Princeton, N.J.: Princeton University Press.

Parson, E. A., and W. C. Clark. 1995. Sustainable development as social learning: Theoretical perspectives and practical challenges for the design of a research program. In Barriers and Bridges to the Renewal of Ecosystems and Institutions, L. Gunderson, C. S. Holling, and S. Light, eds. New York: Columbia University Press.

Patten, D. T. 1991. Glen Canyon Environmental Studies Research Program: Past, present, and future. In Colorado River Ecology and Dam Management. Proceedings of a Symposium, May 24–25, 1990, Santa Fe, N. Mex. Washington, D.C.: National Academy Press.

Patten, D. T. 1993. Long-term monitoring in Glen and Grand Canyon: Response to operations of Glen Canyon Dam. Appendix B in Glen Canyon Environmental Studies: Review of the Draft Federal Long-Term Monitoring Plan for the Colorado River below Glen Canyon Dam. Washington, D.C.: National Research Council.

Patten, D. T. 1998. Integration and Evaluation of Glen Canyon Environmental Studies Research Findings: The Grand Canyon Riverine Ecosystem—Functions, Processes and Relationships Among Biotic and Abiotic Driving and Response Variables. Salt Lake City, Utah: U.S. Bureau of Reclamation, and Flagstaff, Ariz.: Grand Canyon Monitoring and Research Center.

Phillips, A. M. III, and L. Jackson. 1997. Monitoring Hualapai Ethnobotanical Resources Along the Colorado River, 1997. Annual Report. Hualapai Tribe, Cultural Resources Division, Peach Springs, Ariz.

Pisani, D. 1992. To Reclaim a Divided West: Water, Law, and Public Policy, 1848-1902. Albuquerque: University of New Mexico Press.

Pontius, D. 1997. Colorado River Basin Study. Denver, Colo.: Western Water Policy Review Advisory Commission.

Powell, J. W. 1874. Exploration of the Colorado River and its Canyons. New York: Dover.

Prescott, J. R. V. 1987. Political Frontiers and Boundaries. London: Allen and Unwin.

President's Water Resources Policy Commission. 1950. Ten Rivers in America's Future. Vol. 2, pp. 355-462 on the Colorado River. Washington, D.C.: U.S. Government Printing Office.

Pulwarty, R. S., and K. T. Redmond. 1997. Climate and salmon restoration in the Columbia River Basin: The role and usability of seasonal forecasts. Bulletin of the American Meteorological Society 78(3): 381–397.

Pyne, S. 1998. How the Grand Canyon Became Grand. New York: Vintage.

Ralston, B., R. Winfree, and B. Gold. 1998. Beach/Habitat-Building Flow Resource Criteria: A Process Document. Draft report submitted to the Glen Canyon Dam Adaptive Management Program Technical Work Group.

Randle, T. J., R. I. Strand, and A. Streifel. 1993. Engineering and environmental considerations of Grand Canyon sediment management. In Engineering Solutions to Environmental Challenges. 13th Annual USCOLD Lecture, Chattanooga, Tenn. Denver, Colo.: U.S. Committee on Large Dams.

Reitsma, R. F. 1996. Structure and support of water-resources management and decision making. Journal of Hydrology 177:253-268.

Riebsame, W. E., et al. 1997. Atlas of the New American West. New York: W. W. Norton.

Rob, P., and C. Coronel. 1997. Database systems: Design, Implementation, and Management. Cambridge, Mass.: Course Technology.

Roberts, A., R. M. Begay, and K. B. Kelley. 1995. Bits'iis Nineezi (The River of Neverending Life): Navajo History and Cultural Resources of the Grand Canyon and the Colorado River. Window Rock, Ariz.: Navajo Nation Historic Preservation Department.

Rogers, K. 1998. Managing science/management partnerships: A challenge of adaptive management. Conservation Ecology 2(2). [http://www.consecol.org/journal/vol2/iss2/respl.].

Rosenberg, K. V., R. D. Ohmart, W. C. Hunter, and B. W. Anderson. 1991. Birds of the Lower Colorado River Valley. Tucson: University of Arizona Press.

Rossi, P. H., and H. E. Freeman. 1993. Evaluation: A Systematic Approach. 5th ed. Thousand Oaks, Calif.: Sage Publications.

Rubin, D. M., J. M. Nelson, and D. J. Topping. 1998. Relation of inversely graded deposits to suspended-sediment grain-size evolution during the 1996 flood experiment in Grand Canyon. Geology 26(2):99–102.

Ruffner, G. A., N. J. Czaplewski, and S. W. Carothers. 1978. Distribution and natural history of some mammals from the inner gorge of the Grand Canyon, Arizona. Journal of the Arizona–Nevada Academy of

Science 13: 85–91.

Society for American Archaeology Bulletin. 1993. Working Together Series.

Schmidt, J. C. 1999. Summary and synthesis of geomorphic studies conducted during the 1996 controlled flood in Grand Canyon. In The Controlled Flood in Grand Canyon, R. H. Webb, J. C. Schmidt, G. R. Marzolf, and R. A. Valdez, eds. Geophysical Monograph Series Vol. 110. Washington, D.C.: American Geophysical Union.

Schmidt, J. C., and J. B. Graf. 1990. Aggradation and degradation of alluvial sand deposits, 1965 to 1986, Colorado River, Grand Canyon National Park, Arizona. U.S. Geological Survey Professional Paper 1493. Washington, D.C.: U.S. Geological Survey.

Schmidt, J. C., P. E. Grams, and R. H. Webb. 1995. Comparison of the magnitude of erosion along two large regulated rivers. Water Resources Bulletin 31(4):617–631.

Schmidt, J. C., R. H. Webb, R. A. Valdez, G. R. Marzolf, and L. E. Stevens. 1998. Science and Values in River Restoration in the Grand Canyon. BioScience 48(9):735–747.

Schmidt, J. C., E. D. Andrews, D. L. Wegner, D. T. Patten, G. R. Marzolf, and T. O. Moody. 1999a. Origins of the 1996 controlled flood in Grand Canyon. In The Controlled Flood in Grand Canyon, R. H. Webb, J. C. Schmidt, G. R. Marzolf, and R. A. Valdez, eds. Geophysical Monograph Series Vol. 110. Washington, D.C.: American Geophysical Union.

Schmidt, J. C., P. E. Grams, and M. F. Leschin. 1999b. Variation in the Magnitude and Style of Deposition and Erosion in Three Long (8-12 km) Reaches as Determined by Photographic Analyses. In The Controlled Flood in Grand Canyon, R. H. Webb, J. C. Schmidt, G. R. Marzolf, and R. A. Valdez, eds. Geophysical Monograph Series Vol. 110. Washington, D.C.: American Geophysical Union.

Segal-Horn, S., ed. 1998. The Strategy Reader. Oxford: Blackwell.

Shadish, W. R., Jr., T. D. Cook, L. C. Leviton. 1991. Foundations of Program Evaluation: Theories of Practice. Newbury Park, Calif.: Sage Publications.

Sit, V., and B. Taylor, eds. 1998. Statistical methods for adaptive management studies. Land Management Handbook No. 42. Victoria, British Columbia: Ministry of Forests Research Program.

Smillie, G. M., W. L. Jackson, and D. Tucker. 1993. sand budget: Lee's Ferry to Little Colorado River. National Park Service Technical Report NPS/NRWRD/NRTR-92-12. Washington, D.C.: U.S. Department of the Interior.

Smith, J. B. 1997. Setting priorities for adapting to climate change. Global Environmental Change: Human and Policy Dimensions 7:251–264.

Smith, J. D. 1999. Flow and suspended-sediment transport in the Colorado River near National Canyon. In R. H. Webb, J. C. Schmidt, G. R. Marzolf, and R. A. Valdez, eds. The Controlled Flood in Grand Canyon. Geophysical Monograph Series Vol. 110. Washington, D.C.: American Geophysical Union.

Smith, W. E. 1998. The Struggle for Water. Chicago: University of Chicago Press.

Smithers, J., and B. Smit. 1997. Human adaptation to climatic variability and change. Global Environmental Change: Human and Policy Dimensions 7:129–146.

Stevens, L. 1983. The Colorado River in Grand Canyon: A Guide. Flagstaff, Ariz.: Red Lake Books.

Stoffle, R. W., D. E. Austin, and B. K. Fulfrost. 1994. Piapaxa 'Uipi (Big River Canyon). Tucson: University of Arizona, Bureau of Applied Research in Anthropology.

Stoffle, R. W. et al. 1995. Itus, Auv, Te'ek (Past, present, future). Managing Southern Paiute Resources in the Colorado River Corridor. Pipe Spring, Ariz. Southern Paiute Consortium. Tucson: University of Arizona, Bureau of Applied Research in Anthropology.

Suttkus, R. D., and G. H. Clemmer. 1979. The fishes of the Colorado River in Grand Canyon National Park. In Proceedings of the First Conference on Scientific Research in the National Parks, New Orleans, November 9–12, 1976. U.S. National Park Service Transactions Proceedings Series 1(5):599–604.

S. W. Carothers and Associates. 1998. Cultural resources synthesis contract. Proposal and draft report on file with the National Research Council, Washington, D.C.

Tarlock, D. 1996. Environmental law: Ethics or science? In Beyond the Balance of Nature: Environmental Law Faces the New Ecology. First annual Cummings Colloquium on Environmental Law. Duke Environmental Law and Policy Forum 7:193–223.

Taylor, B., L. Kremsater, and R. Ellis. 1997. Adaptive management of forests in British Columbia. Victoria, British Columbia: Ministry of Forests.

Technical Work Group. 1997. Report of the Spike Flow Subgroup. October, 1997.

Technical Work Group. 1999. Draft Guidance Document outline. Photocopy on file with National Research Council, Washington, D.C.

Technical Work Group Minutes. 1997. http://www.uc.usbr.gov/amp/twg/mtg_1197.html.

Topping, D. J., et al. 1999. Linkage of grain-size evolution and sediment depletion during Colorado River floods. In The Controlled Flood in Grand Canyon, R. H. Webb, J. C. Schmidt, G. R. Marzolf, and R. A. Valdez, eds. Geophysical Monograph Series Vol. 110. Washington, D.C.: American Geophysical Union.

Tsosie, R. 1998. Presentation for Glen Canyon Dam Technical Work Group: Federal agencies and the trust responsibilities. Unpublished paper on file with the National Research Council, Washington, D.C.

Turner, R. M., and M. M. Karpiscak. 1980. Recent Vegetation Changes Along the Colorado River Between Glen Canyon Dam and Lake Mead, Arizona. U.S. Geological Survey Professional Paper 1132.

U. S. Bureau of Reclamation. 1950. Colorado River Storage Project and Participating Projects Upper Colorado River Basin. Project Planning Report No. 4-81.81-1. Salt Lake City, Utah: U.S. Bureau of Reclamation.

U.S. Bureau of Reclamation. 1990. Glen Canyon Dam Environmental Impact Statement, Final Analysis Report on Scoping Comments. Prepared by Bear West Consulting Team for Bureau of Reclamation. Salt Lake City, Utah.

U.S. Bureau of Reclamation. 1995. Operation of Glen Canyon Dam: Final Environmental Impact Statement. Washington, D.C.: U.S. Government Printing Office.

U.S. Bureau of Reclamation. 1999. Glen Canyon Dam modifications to control downstream temperatures. Plan and Draft Environmental Impact Assessment. Salt Lake City, Utah: U.S. Bureau of Reclamation.

U.S. Congress, House of Representatives. Subcommittee on National Parks and Public Lands and Subcommittee on Water and Power. 1997. Joint hearing on the Sierra Club's proposal to drain Lake Powell or reduce its water storage capability. September 24, 1997. 105th Cong., 1st sess. H. Rpt. 105–156.

U.S. Department of the Interior. 1946. The Colorado River: A Natural Menace Becomes a National Resource. Washington, D.C.: U.S. Government Printing Office.

U.S. Geological Survey. 1925. Water Power and Flood Control of the Colorado River below Green River, Utah. Water Supply Paper 556. Washington, D.C.: U.S. Government Printing Office.

U.S. National Park Service. 1946. A Survey of the Recreational Resources of the Colorado River Basin. Washington, D. C.: U.S.

Government Printing Office.

Valdez, R. A., and S. W. Carothers. 1998. The aquatic ecosystem of the Colorado River in Grand Canyon. Grand Canyon Data Integration Project Synthesis Report prepared for the Bureau of Reclamation. ·

Vernieu, W. S., and S. J. Hueftle. 1999. Integrated Water Quality Program. Flagstaff, Ariz.: Grand Canyon Monitoring and Research Center.

Volkman, J. M. 1997. A River in Common: The Columbia River, the Salmon Ecosystem, and Water Policy. Denver, Colo.: Western Water Policy Review Advisory Commission.

Volkman, J. M., and W. E. McConnaha. 1993. Through a glass, darkly: Columbia River salmon, the Endangered Species Act and adaptive management. Environmental Law 23:1249–1272.

Walters, C. J. 1986. Adaptive Management of Natural Resources. New York: McGraw Hill.

Walters, C. J. 1997. Challenges in adaptive management of riparian and coastal ecosystems. Conservation Ecology 1:3, pp. 1ff. [http://www.consecol.org/vol1/iss2/art1].

Walters, C. J. 1998. Grand Canyon conceptual modeling workshop. Saguaro Lake, Ariz.

Walters, C. J., and C. S. Holling. 1990. Large-scale management experiments and learning by doing. Ecology 71: 2060–2068.

Walters, J. C., L. Gunderson, and C. S. Holling. 1992. Experimental policies for water management in the Everglades. Ecological Applications 2:189–202.

Webb, R. H. 1996. Grand Canyon, a Century of Change: Rephotography of the 1889–1890 Stanton Expedition. Tucson: University of Arizona Press.

Wegner, D. L. 1991. Brief history of the Glen Canyon Environmental Studies. Pp. 226–238 in Colorado River Ecology and Dam Management. Proceedings of a Symposium, May 24–25, 1990, Santa Fe, N. Mex. Washington, D.C.: National Academy Press.

Wescoat, J. L., Jr. 1987. The practical range of choice in water resource geography. Progress in Human Geography 11:41–59.

Wescoat, J. L., Jr. 1992. Common themes in the work of Gilbert White and John Dewey: A pragmatic appraisal. Annals of the Association of American Geographers 82:587–607.

Westley, F. 1995. Governing design. In Barriers and Bridges to the Renewal of Ecosystems and Institutions. L. Gunderson, C. S. Holling, and S. Light, eds. New York: Columbia University Press.

White, G. F. 1957. A Perspective of River Basin Development. Law and

Contemporary Problems 22 (Spring 1957):157–184.

Wiele, S. M., E. D. Andrews, and E. R. Griffin. 1999. The effect of sand concentration on depositional rate, magnitude, and location in the Colorado River below the Little Colorado River. In The Controlled Flood in Grand Canyon, R. H. Webb, J. C. Schmidt, G. R. Marzolf, and R. A. Valdez, eds. Geophysical Monograph Series Vol. 110. Washington, D.C.: American Geophysical Union.

Wohl, E. 1998. Preliminary Report of the Physical Resources Monitoring Review Panel, Submitted to Grand Canyon Monitoring and Research Center (September 1998).

Appendix A

RECLAMATION PROJECTS AUTHORIZATION AND ADJUSTMENT
ACT OF 1992

TITLE XVIII — GRAND CANYON PROTECTION

SECTION 1801. SHORT TITLE

This Act may be cited as the
"Grand Canyon Protection Act of 1992."

SEC. 1802. PROTECTION OF GRAND CANYON NATIONAL PARK.

(a) In General. — The Secretary shall operate Glen Canyon Dam in accordance with the additional criteria and operating plans specified in section 1804 and exercise other authorities under existing law in such a manner as to protect, mitigate adverse impacts to, and improve the values for which Grand Canyon National Park and Glen Canyon National Recreation Area were established, including, but not limited to natural and cultural resources and visitor use.

(b) Compliance With Existing Law. — The Secretary shall implement this section in a manner fully consistent with and subject to the Colorado River Compact, the Upper Colorado River Basin Compact, the Water Treaty of 1944 with Mexico, the decree of the Supreme Court in Arizona v. California, and the provisions of the Colorado River Storage Project Act of

1956 and the Colorado River Basin Project Act of 1968 that govern allocation, appropriation, development, and exportation of the waters of the Colorado River basin.

(c) Rule of Construction. — Nothing in this title alters the purposes for which the Grand Canyon National Park or the Glen Canyon National Recreation Area were established or affects the authority and responsibility of the Secretary with respect to the management and administration of the Grand Canyon National Park and Glen Canyon National Recreation Area, including natural and cultural resources and visitor use, under laws applicable to those areas, including, but not limited to, the Act of August 25, 1916 (39 Stat. 535) as amended and supplemented.

SEC. 1803. INTERIM PROTECTION OF GRAND CANYON NATIONAL PARK.

(a) Interim Operations. — Pending compliance by the Secretary with section 1804, the Secretary shall, on an interim basis, continue to operate Glen Canyon Dam under the Secretary's announced interim operating criteria and the Interagency Agreement between the Bureau of Reclamation and the Western Area Power Administration executed October 2, 1991 and exercise other authorities under existing law, in accordance with the standards set forth in Section 1802, utilizing the best and most recent scientific data available.

(b) Consultation. — The Secretary shall continue to implement Interim Operations in consultation with—

> (1) Appropriate agencies of the Department of the Interior, including the Bureau of Reclamation, United States Fish and Wildlife Service, and the National Park Service;
> (2) The Secretary of Energy;
> (3) The Governors of the States of Arizona, California, Colorado, Nevada, New Mexico, Utah, and Wyoming;
> (4) Indian Tribes; and
> (5) The general public, including representatives of the academic and scientific communities, environmental organizations, the recreation industry, and contractors for the purchase of Federal power produced at Glen Canyon Dam.

(c) Deviation From Interim Operations. — The Secretary may deviate from Interim Operations upon a finding that deviation is necessary and in the public interest to —

(1) comply with the requirements of Section 1804(a);
(2) respond to hydrologic extremes or power system operation emergencies;
(3) comply with the standards set forth in Section 1802;
(4) respond to advances in scientific data; or
(5) comply with the terms of the Interagency Agreement.

(d) Termination of Interim Operations. — Interim operations described in this section shall terminate upon compliance by the Secretary with Section 1804.

SEC. 1804.GLEN CANYON DAM ENVIRONMENTAL IMPACT STATEMENT; LONG-TERM OPERATION OF GLEN CANYON DAM.

(a) Final Environmental Impact Statement. — Not later than 2 years after the date of enactment of this Act, the Secretary shall complete a final Glen Canyon Dam environmental impact statement, in accordance with the National Environmental Policy Act of 1969 (42 U.S.C. 4321 et. seq.).

(b) Audit. — The Comptroller General shall—

(1) audit the costs and benefits to water and power users and to natural, recreational, and cultural resources resulting from management policies and dam operations identified pursuant to the environmental impact statement described in subsection (a); and
(2) report the results of the audit to the Secretary and the Congress.

(c) Adoption of Criteria and Plans. —

(1) Based on the findings, conclusions, and recommendations made in the environmental impact statement prepared pursuant to subsection (a) and the audit performed pursuant to subsection (b),

the Secretary shall —

> (A)   adopt criteria and operating plans separate from and in addition to those specified in section 602(b) of the Colorado River Basin Project Act of 1968 and
> (B)   exercise other authorities under existing law, so as to ensure that Glen Canyon Dam is operated in a manner consistent with section 1802.

(2) Each year after the date of the adoption of criteria and operating plans pursuant to paragraph (1), the Secretary shall transmit to the Congress and to the Governors of the Colorado River Basin States a report, separate from and in addition to the report specified in section 602(b) of the Colorado River Basin Project Act of 1968 on the preceding year and the projected year operations undertaken pursuant to this Act.

(3) In preparing the criteria and operating plans described in section 602(b) of the Colorado River Basin Project Act of 1968 and in this subsection, the Secretary shall consult with the Governors of the Colorado River Basin States and with the general public, including—

> (A)   representatives of academic and scientific communities;
> (B)   environmental organizations;
> (C)   the recreation industry; and
> (D)   contractors for the purchase of Federal power produced at Glen Canyon Dam.

(d) Report to Congress. — Upon implementation of long-term operations under subsection (c), the Secretary shall submit to the Congress the environmental impact statement described in subsection (a) and a report describing the long-term operations and other reasonable mitigation measures taken to protect, mitigate adverse impacts to, and improve the condition of the natural recreational, and cultural resources of the Colorado River downstream of Glen Canyon Dam.

(e) Allocation of Costs. — The Secretary of the Interior, in consultation with the Secretary of Energy, is directed to reallocate the costs of construction, operation, maintenance, replacement and emergency expenditures for Glen Canyon Dam among the purposes directed in section 1802 of this Act and the purposes established in the Colorado River Storage Project Act of April 11, 1956 (70 Stat. 170). Costs allocated to section 1802 purposes shall be nonreimbursable. Except that in Fiscal Year 1993 through 1997 such costs shall be nonreimbursable only to the extent to which the Secretary finds the effect of all provisions of this Act is to increase net offsetting receipts; Provided, further that if the Secretary finds in any such year that the enactment of this Act does cause a reduction net offsetting receipts generated by all provisions of this Act, the costs allocated to section 1802 purposes shall remain nonreimbursable. The Secretary shall determine the effect of all the provisions of this Act and submit a report to the appropriate House and Senate committees by January 31 of each fiscal year, and such report shall contain for that fiscal year a detailed accounting of expenditures incurred pursuant to this Act, offsetting receipts generated by this Act, and nay increase or reduction in net offsetting receipts generated by this Act.

SEC. 1805. LONG-TERM MONITORING

(a) In General. — The Secretary shall establish and implement long-term monitoring programs and activities that will ensure that Glen Canyon Dam is operated in a manner consistent with that of section 1802.

(b) Research. — Long-term monitoring of Glen Canyon Dam shall include any necessary research and studies to determine the effect of the Secretary's actions under section 1804(c) on the natural, recreational, and cultural resources of Grand Canyon National Park and Glen Canyon National Recreation Area.

(c) Consultation. — The monitoring programs and activities conducted under subsection (a) shall be established and implemented in consultation with—

> (1) the Secretary of Energy;
> (2) the Governors of the States of Arizona, California, Colorado, Nevada, New Mexico, Utah, and Wyoming;
> (3) Indian tribes; and
> (4) the general public, including representatives of academic and scientific communities, environmental organizations, the recrea-

tion industry, and contractors for the purchase of Federal power produced at Glen Canyon Dam.

SEC. 1806. RULES OF CONSTRUCTION

Nothing in this title is intended to affect in any way—

(1) the allocations of water secured to the Colorado Basin States by any compact, law, or decree; or
(2) any Federal environmental law, including the Endangered Species Act (16 U.S.C. 1531 et seq.).

SEC. 1807. STUDIES NONREIMBURSABLE

All costs of preparing the environmental impact statement described in section 1804, including supporting studies, and the long-term monitoring programs and activities described in section 1805 shall be nonreimbursable. The Secretary is authorized to use funds received from the sale of electric power and energy from the Colorado River Storage Project to prepare the environmental impact statement described in section 1804, including supporting studies, and the long-term monitoring programs and activities described in section 1805, except that such funds will be treated as having been repaid and returned to the general fund of the Treasury as costs assigned to power for repayment under section 5 of the Act of April 11, 1956 (70 Stat. 170). Except that in Fiscal Year 1993 through 1997 such provisions shall take effect only to the extent to which the Secretary finds the effect of all the provisions of this Act is to increase net offsetting receipts; Provided, further that if the Secretary finds in any such year that the enactment of this Act does cause a reduction in net offsetting receipts generated by all provisions of this Act, all costs described in this section shall remain nonreimbursable. The Secretary shall determine the effect of all the provisions of this Act and submit a report to the appropriate House and Senate committees by January 31 of each fiscal year, and such report shall contain for that fiscal year a detailed accounting of expenditures incurred pursuant to this Act, offsetting receipts generated by this Act, and any increase or reduction in net offsetting receipts generated by this Act.

SEC. 1808. AUTHORIZATION OF APPROPRIATIONS.

There are authorized to be appropriated such sums as are necessary to carry out this title.

SEC. 1809. REPLACEMENT POWER.

The Secretary of Energy in consultation with the Secretary of the Interior and with epresentatives of the Colorado River Storage Project power customers, environmental organizations and the States of Arizona, California, Colorado, Nevada, New Mexico, Utah and Wyoming shall identify economically and technically feasible methods of replacing any power generation that is lost through adoption of long-term operational criteria for Glen Canyon Dam as required by Section 1804 of this title. The Secretary shall present a report of the findings, and implementing draft legislation, if necessary, not later than two years after adoption of long-term operating criteria. The Secretary shall include an investigation of the feasibility of adjusting operations at Hoover Dam to replace all or part of such lost generation. The Secretary shall include an investigation of the modifications or additions to the transmission system that may be required to acquire and deliver replacement power.

Appendix B

Adaptive Management Work Group (AMWG)

AMWG COMMITTEE MEMBERSHIP

Cooperating Agencies (12)
U.S. Bureau of Indian Affairs
U.S. Bureau of Reclamation
U.S. Fish and Wildlife
    Service
U.S. National Park Service
Western Area Power
    Administration
Arizona Game and Fish
    Department
Hopi Tribe
Hualapai Tribe
Navajo Nation
San Juan Southern Paiute
    Tribe
Southern Paiute Consortium
Pueblo of Zuni

Basin States (7)
Arizona
California
Colorado
Nevada
New Mexico
Wyoming
Utah

Environmental Groups (2)
American Rivers
Grand Canyon Trust

Recreation Interests (2)
Arizona Flycasters/Trout
    Unlimited
Grand Canyon River Guides

**Federal Power Purchase
Contractors (2)**
Colorado River Energy Dist.
    Assoc.
Utah Associated Municipal
    Power

Appendix C

Technical Work Group (TWG)

TWG COMMITTEE MEMBERSHIP

Cooperating Agencies (12)
U.S. Bureau of Indian Affairs
U.S. Bureau of Reclamation
U.S. Fish and Wildlife
Service
U.S. Geological Survey
U.S. National Park Service
Western Area Power
Administration
Arizona Game and Fish
Department
Hopi Tribe
Hualapai Tribe
Navajo Nation
Southern Paiute Consortium
Pueblo of Zuni

Basin States (7)
Arizona
California
Colorado
Nevada
New Mexico
Utah
Wyoming

Environmental Groups (2)
American Rivers
Grand Canyon Trust

Recreation Interests (2)
Arizona Flycasters/Trout
Unlimited
Grand Canyon River Guides

Federal Power Purchase Contractors (2)
Colorado River Energy Dist.
Assoc.
Utah Associated Municipal
Power

Appendix D

Record of Decision (1996)

United States Department of the Interior

BUREAU OF RECLAMATION
Upper Colorado Regional Office
125 South State Street, Room 6107
Salt Lake City, Utah 84138-1102

IN REPLY REFER TO

UC-326
ENV-6.00

OCT 2 5 1996

To: All on the Enclosed List

Subject: Record of Decision (ROD) on the Operation of Glen Canyon Dam Final
         Environmental Impact Statement (EIS)

We are pleased to provide you with the enclosed copy of the subject ROD which was
signed by the Secretary of the Interior on October 9, 1996. We are deeply appreciative
of your participation in this long, but highly successful process.

You will note that only two of the attachments to the ROD are included in the enclosed
copy. Attachment 3, the GCES Non-Use Values Final Study Summary Report, is
available on the Internet at **http://www.uc.usbr.gov** which is the Upper Colorado
Region's home page. Attachment 4, the General Accounting Office's Final Audit
Report (GAO/RCED-97-12) may be obtained by calling (202) 512-6000, or on the
Internet at **http://www.gao.gov** which is the General Accounting Office's home page.

If you have any questions about the ROD, please contact either Bruce Moore at
(801) 5245415, or Gordon Lind at (801) 524-3216.

Sincerely,

Charles A. Calhoun
Regional Director

Enclosure

RECORD OF DECISION

OPERATION OF GLEN CANYON DAM

Final Environmental Impact Statement

October 1996

Approved _____

Date **OCT 0 8 1996**

Commissioner, U.S. **Bureau** of Reclamation

_____

Secretary of the Interior

Date **OCT 0 9 1996**

RECORD OF DECISION

OPERATION OF GLEN CANYON DAM
FINAL ENVIRONMENTAL IMPACT STATEMENT

I. INTRODUCTION

This record of decision (ROD) of the Department of the Interior, Bureau of Reclamation (Reclamation), documents the selection of operating criteria for Glen Canyon Dam, as analyzed in the final Environmental Impact Statement (EIS), dated March 21, 1995 (FES 95-8). The EIS on the operation of Glen Canyon Dam was prepared with an unprecedented amount of scientific research, public involvement, and stakeholder cooperation. Scientific evidence gathered during Phase I of the Glen Canyon Environmental Studies (GCES) indicated that significant impacts on downstream resources were occurring due to the operation of Glen Canyon Dam. These findings led to a July 1989 decision by the Secretary of the Interior for Reclamation to prepare an EIS to reevaluate dam operations. The purpose of the reevaluation was to determine specific options that could be implemented to minimize, consistent with law, adverse impacts on the downstream environment and cultural resources, as well as Native American interests in Glen and Grand Canyons. Analysis of an array of reasonable alternatives was needed to allow the Secretary to balance competing interests and to meet statutory responsibilities for protecting downstream resources and producing hydropower, and to protect affected Native American interests, in addition, the Grand Canyon Protection Act of 1992 was enacted on October 30, 1992. Section 1802 (a) of the Act requires the Secretary to operate Glen Canyon Dam:

> "...in such a manner as to protect, mitigate adverse impacts
> to, and improve the values for which Grand Canyon National Park
> and Glen Canyon National Recreation Area were established,
> including, but not limited to natural and cultural resources and
> visitor use. "

Alternatives considered include the No Action Alternative as well as eight operational alternatives that provide various degrees of protection for downstream resources and hydropower production.

II. DECISION

The Secretary's decision is to implement the Modified Low Fluctuating Flow Alternative (the preferred alternative) as described in the final EIS on the Operation of Glen Canyon Dam with a minor change in the timing of beach/habitat building flows (described below). This alternative was selected because it will reduce daily flow fluctuations well below the no action levels (historic pattern of releases) and will provide high steady releases of short duration, which will protect or enhance downstream resources while allowing limited flexibility for power operations.

The Modified Low Fluctuating Flow Alternative incorporates beach/habitat-building flows which are scheduled high releases of short duration designed to rebuild high elevation sandbars, deposit nutrients, restore backwater channels, and provide some of the dynamics of a natural system. In the final EIS, it was assumed that these flows would occur in the spring when the reservoir is low, with a frequency of 1 in 5 years.

The Basin States expressed concern over the beach/habitat-building flows described in the final EIS because of the timing of power plant bypasses. We have accommodated their concerns, while maintaining the objectives of the beach/habitat-building flows. Instead of conducting these flows in years in which Lake Powell storage is low on January 1, they will be accomplished by utilizing reservoir releases in excess of power plant capacity required for dam safety purposes. Such releases are consistent with the 1956 Colorado River Storage Project Act, the 1968 Colorado River Basin Project Act, and the 1992 Grand Canyon Protection Act.

Both the Colorado River Management Work Group and the Transition Work Group, which participated in the development of the Annual Operating Plan and the EIS, respectively, support this change as it conforms unambiguously with each member's understanding of the Law of the River. These groups include representatives of virtually all stakeholders in this process. The upramp rate and maximum flow criteria were also modified between the draft and final EIS. The upramp rate was increased from 2,500 cubic feet per second per hour to 4,000 cubic feet per second per hour, and the maximum allowable release was increased from 20,000 to 25,000 cubic feet per second. We made these modifications to enhance power production flexibility, as suggested by comments received. These modifications were controversial among certain interest groups because of concerns regarding potential impacts on resources in the

Colorado River and the Grand Canyon. However, our analysis indicates that there would be no significant differences in impacts associated with these changes ("Assessment of Changes to the Glen Canyon Dam EIS Preferred Alternative from Draft to Final EIS", October 1995).

The 4,000 cubic feet per second per hour upramp rate limit will be implemented with the understanding that results from the monitoring program will be carefully considered. If impacts differing from those described in the final EIS are identified, a new ramp rate criterion will be considered by the Adaptive Management Work Group and a recommendation for action forwarded to the Secretary.

The maximum flow criterion of 25,000 cubic feet per second will be implemented with the understanding that actual maximum daily releases would only occasionally exceed 20,000 cubic feet per second during a minimum release year of 8.23 million acre-feet. This is because the maximum allowable daily change constraint overrides the maximum allowable release and because monthly release volumes are lower during minimum release years. If impacts differing from those described in the final EIS are identified through the Adaptive Management Program, the maximum flow restriction will be reviewed by the Adaptive Management Work Group and a recommendation for action will be forwarded to the Secretary.

III. DESCRIPTION OF ALTERNATIVES

Nine alternative methods of operating Glen Canyon Dam (including the No Action Alternative) were presented in the final EIS. The eight action alternatives were designed to provide a reasonable range of alternatives with respect to operation of the dam. One alternative would allow unrestricted fluctuations in flow (within the physical constraints of the power plant) to maximize power production, four would impose varying restrictions on fluctuations, and three others would provide steady flows on a monthly, seasonal, or annual basis. The names of the alternatives reflect the various operational regimes. In addition, the restricted fluctuating flow and steady flow alternatives each include seven elements which are common to all of them. These common elements are: 1) Adaptive Management, 2) Monitoring and Protecting Cultural Resources, 3) Flood Frequency Reduction Measures, 4) Beach/Habitat-Building Flows, 5) New Population of Humpback Chub, 6) Further Study of Selective Withdrawal, and 7) Emergency Exception Criteria. A detailed description of the aitematives and common elements can be found in Chapter 2 of the final EIS. A brief description of the alternatives is given below.

UNRESTRICTED FLUCTUATING FLOWS

No Action: Maintain the historic pattern of fluctuating releases up to 3 1,500 cubic feet per second and provide a baseline for impact comparison.

Maximum Power plant Capacity: Permit use of full power plant capacity up to 33,200 cubic feet per second.

RESTRICTED FLUCTUATING FLOWS

High: Slightly reduce daily fluctuations from historic levels.

Moderate: Moderately reduce daily fluctuations from historic levels; includes habitat maintenance flows.

Modified Low (Preferred Alternative): Substantially' reduce daily fluctuations from historic levels; includes habitat maintenance flows.

Interim Low: Substantially reduce daily fluctuations from historic levels; same as interim operations except for addition of common elements.

STEADY FLOWS

Existing Monthly Volume: Provide steady flows that use historic monthly release strategies.

Seasonally Adjusted: Provide steady flows on a seasonal or monthly basis; includes habitat maintenance flows.

Year-Round: Provide steady flows throughout the year.
Table 1 shows the specific operational criteria for each of the alternatives.

IV. SIGNIFICANT ISSUES AND ALTERNATIVES

The Glen Canyon Dam EIS scoping process was initiated in early 1990 and the public was invited to comment on the appropriate scope of the EIS. More than 17,000 comments were received during the scoping period, reflecting the national attention and intense interest in the EIS.

As a result of the analysis of the oral and written scoping comments, the following were determined to be resources or issues of public concern: beaches, endangered species, ecosystem integrity, fish, power costs, power production, sediment, water conservation, rafting/boating, air quality, the Grand Canyon wilderness, and a category designated as "other" for remaining concerns. Comments regarding interests and values were categorized as: expressions about the Grand Canyon, economics, nonquantifiable values, nature versus human use, and the complexity of Glen Canyon Dam issues.

The EIS team consolidated and refined the public issues of concern, identifying the significant resources and associated issues to be analyzed in detail. These resources include: water, sediment, fish, vegetation, wildlife and habitat, endangered and other special status species, cultural resources, air quality, recreation, hydropower, and non-use value.

Further meetings were held with representatives from the cooperating agencies and public interest groups who provided comments on the criteria for development of reasonable alternatives for the EIS. The public also had an opportunity to comment on the preliminary selection of alternatives at public meetings and through mailings. The final selection of alternatives took into consideration the public's views.

V. COMMENTS RECEIVED ON THE FINAL EIS

Many comments and recommendations on the final EIS were received in the form of pre-printed postcards and letters that addressed essentially the same issues. The comments are summarized below along with Reclamation's responses.

COMMENT: Maintain Draft EIS flows. Modifying the upramp rate and maximum flows between the draft and final EIS has neither been open for public review nor subjected to serious scientific scrutiny. These changes should have been addressed in the draft EIS and made available for public

Table 1.-Operating limits of alternatives identified for detailed analysis

	Unrestricted Fluctuating Flows		Restricted Fluctuating Flows				Steady Flows		
	No Action	Maximum Powerplant Capacity	High	Moderate	Modified Low	Interim Low	Existing Monthly Volume	Seasonally Adjusted	Year-Round
Minimum Releases (cfs)	1,000 Labor Day Easter [3]3,000 Easter Labor Day	1,000 Labor Day Easter [2]3,000 Easter Labor Day	3,000 5,000 8,000 depending on monthly volume, firmload, and market conditions	5,000	8,000 between 7 a.m. and 7 p.m. 5,000 at night	[3]8,000 between 7 a.m. and 7 p.m. 5,000 at night	8,000	8,000 Oct Nov 8,500 Dec 11,000 Jan Mar 12,500 Apr 18,000 May Jun 12,500 Jul 9,000 Aug Sep	Yearly Volume Prorated *
Maximum Releases (cfs)[5]	31,500	33,200	31,500	31,500 (may be exceeded during habitat maintenance flows)	25,000 (exceeded during habitat maintenance flows)	20,000	Monthly Volumes Prorated	18,000 (exceeded during habitat maintenance flows)	Yearly volume Prorated
Allowable daily flow fluctuations (cfs/24 hours)	30,500 Labor Day-Easter 28,500 Easter Labor Day	32,200 Labor Day-Easter 30,200 Easter – Labor Day	15,000 to 22,000	±45% of mean flow for the month not to exceed ±6000	[4]5,000, 6,000 or 8,000	[6]5,000, 6,000 or 8,000	[7]±1000	[7]±1000	[7]±1000

Ramp Rates (cfs/hour)	Unrestricted	Unrestricted	Unrestricted up 5,000 or 4,000 down	4,000 up 2,500 down	4,000 up 1,500 down	2,500 up 1,500 down	2,000 cfs/day between months	2,000 cfs/day between months	2,000 cfs/day between months
Common Elements	None	None	Adaptive management (including long-term monitoring and research) Monitoring and protecting cultural resources Flood frequency reduction measures Beach/habitat building flows New population of humpback chub Further study of selective withdrawal Emergency exception criteria						

[1] In high volume release months, the allowable daily change would require higher minimum flows

[2] Releases each weekday during recreation season (Easter to Labor Day) would average not less than 8,000 cfs for the period from 8:00 a.m. to midnight

[3] Based on an 8.23-million-acre-foot (maf) year; in higher release years, additional waters would be added equally to each month, subject to a 18,000-cfs maximum.

[4] For an 8.23-maf year, steady flow would be about 11,400 cfs.

[5] Maximums represent normal or routine limits and may necessarily be exceeded during higher water level years.

[6] Daily fluctuation limit of 5,000 cfs for daily release volumes less than 600,000 acre-feet; 6,000 cfs for monthly release volumes of 600,000 to 800,000 acre-feet; and 8,000 cfs for monthly volumes over 800,000 acre-feet.

[7] Adjustments would allow for small power system load changes.

comment at that time. Credible proof, based on the testing of a specific scientific hypothesis, that alterations in operating procedures at Glen Canyon Dam follow the spirit and intent of the Grand Canyon Protection Act needs to be provided. The burden of proof that there will be no impact on downstream resources rests with those proposing changes.

RESPONSE: The modification of the preferred alternative, which incorporated changes in the upramp rate and maximum flows, was made after extensive public discussion. The new preferred alternative was discussed as an agenda item during the May, June, August, and November 1994 public meetings of the Cooperating Agencies who assisted in the development of the EIS. A wide range of public interest groups received advance mailings and agendas and were represented at the public meetings. The environmental groups attending these meetings included: America Outdoors, American Rivers, Desert Flycasters, Environmental Defense Fund, Friends of the River, Grand Canyon River Guides, Grand Canyon Trust, Sierra Club, and Trout Unlimited. Meeting logs indicate that representatives from at least some of these groups attended all but the May meeting. In addition, approximately 16,000 citizens received periodic newsletters throughout the EIS process. This included a newsletter outlining the proposed changes issued several months prior to the final EIS. The environmental groups mentioned above were included on the newsletter mailing list.

Reclamation's research and analysis has been thorough with regards to changes in flows and ramping rates and potential impacts upon downstream resources. A complete range of research flows was conducted from June 1990 to July 1991. These included high and low fluctuating flows with fast and slow up and down ramp rates. Glen Canyon Environmental Studies Phase II identified cause and effect relationships between downramp rates and adverse impacts to canyon resources. However, no cause and effect relationships between upramp rates and adverse impacts to canyon resources were identified. The draft EIS, (a public document peer reviewed by GCES and the EIS Cooperating Agencies) states that upramp rates have not been linked to sandbar erosion (page 95) and that "Rapid increases in river stage would have little or no effect on sandbars." (page 190).

With respect to potential impacts occurring with the change in flows, it should be noted that sand in the Grand Canyon is transported almost exclusively by river flows. The amount of sand transported increases exponentially with increases in river flow. Maintaining sandbars over the long term depends on the amount of sand supplied by tributaries, monthly release volumes, range of flow fluctuations, and the frequency and distribution of flood flows. Conversely, occasional flows between 20,000 and 25,000 cubic feet per second may cause minor beach building, and may provide water to riparian vegetation.

As part of the EIS, the effects of each alternative on long-term sand storage in Marble Canyon (river miles 0 to 61) were analyzed. The Marble Canyon reach was chosen for analysis because it is more sensitive to impacts from dam operations than downstream reaches. For each fluctuating flow alternative, the analysis used 20 years of hourly flow modeled by Spreck Rosekrans of the Environmental Defense Fund and 85 different hydrologic scenarios (each representing 50 years of monthly flow data). This analysis was documented in the draft EIS on page 182, and Appendix D, pages 4-5. The analyses relating to the probability of net gain in riverbed sand for each alternative is documented in the draft EIS on pages 54-55, 184, 187, and 194.

Specific peer reviewed studies relating to the above analyses are listed in Attachment 1.

COMMENT: Do not change the upramp rate and maximum flow criteria at the same time. While acknowledging Reclamation's good efforts to identify and establish optimum operating criteria for all users of Glen Canyon Dam, changing two flow criteria (upramp rate and maximum flow criterion of preferred alternative) does not make prudent scientific sense. It will not result in reliable data. Not enough information is at hand to predict the outcome of these proposals.

RESPONSE: Viewed from the purely scientific viewpoint, it would be preferable to change variables one at a time in a controlled experiment. However, many uncontrolled variables already exist, and from a resource management standpoint the interest lies in measuring the possible resource impact, if any, which might result from jointly changing both criteria. The best available information suggests that the long-term impact of changing both criteria at once will be difficult, if not impossible to detect.

Even though both parameters would change, for 8 months of an 8.23 million-acre foot year (minimum release year), only the upramp rate will be used. The ability to operationally exceed 20,000 cubic feet per second only exists in months in which releases are in excess of 900,000 acre feet. In a minimum release year, flows above 20,000 cubic feet per second will most likely occur in December, January, July, and August.

Evaluation of the upramp rates can be initiated immediately with the evaluation of the increase in maximum flow relegated to the months with the highest volumes. New upramp and maximum flow criteria would be recommended through the Adaptive Management Program should monitoring results indicate that either of these criteria are resulting in adverse impacts to the natural, cultural, or recreational (human safety) resources of the Grand Canyon differing from those shown in the final EIS.

COMMENT: "Habitat/Beach Building Floods" designed to redeposit sediment and reshape the river's topography much like the Canyon's historic floods should be conducted. An experimental release based on this premise is critical to restore some of the river's historic dynamics; without it, any flow regime will result in continued loss of beach and backwater habitat. This "spike" should be assessed and implemented for the spring of 1996, subject to a critical evaluation of its flow size, timing, impact on fisheries, and completion of a comprehensive monitoring plan. Recent side-canyon floods underscore the need for restoring natural processes.

RESPONSE: Reclamation and the Cooperating Agencies continue to support this concept. The preferred alternative supports such a flow regime. A test flow was conducted this spring. The results of this flow are currently being analyzed. We expect to conduct more of these flows in the future.

COMMENT: Endorse the Fish & Wildlife Service's Biological Opinion and implement experimental steady flows to benefit native fishes, subject to the results of a risk/benefit analysis now in progress.

RESPONSE: The preferred alternative provides for experimental steady flows through the Adaptive Management Program for the reasons put forth in the Biological Opinion.

COMMENT: Fund and implement immediately an Adaptive Management Program. This is the appropriate forum to address important issues. It is imperative that resource management relies on good science to monitor, and respond to possible adverse effects resulting from changes in dam operations.

RESPONSE: The preferred alternative provides for implementation of an Adaptive Management Program.

COMMENT: Interior Secretary Babbitt should issue a Record of Decision by December 31, 1995, and conduct an efficient and timely audit by the General Accounting Office as mandated by the Grand Canyon Protection Act.

RESPONSE: In compliance with the Grand Canyon Protection Act, Interior Secretary Babbitt could not issue the Record of Decision until considering the findings of the General Accounting Office. Those findings were issued on October 2, 1996.

OTHER COMMENTS: Another set of comments was received from municipalities and other power user groups. These letters made up about 3 percent of the total received and were essentially identical in content. Although the authors were not totally in agreement with the preferred alternative because of the reduction in peaking power, they believe it is a workable compromise. These letters characterized the final EIS as 'I. . .a model for resolving complex environmental issues among divergent interests." They also urged the government to protect the integrity of the process, resist efforts to overturn the FEIS, and allow the scientist's assessment to stand, in as much **as** the **Adaptive** Management Process will give Reclamation an opportunity to evaluate the effects of operational changes over time and make modifications according to scientific findings.

RESPONSE: While the preferred alternative may not satisfy all interests, Reclamation believes it is a workable compromise and meets the two criteria set out in the EIS for the reoperation of the dam, namely restoring downstream resources and maintaining hydropower capability and flexibility.

A letter of comment from the Environmental Protection Agency (EPA) indicates that EPA's comments on the draft EIS were adequately addressed in the final EIS. It also expresses their support for the preferred alternative.

Samples of the comment letters and cards, and a copy of EPA's comment letter are included as Attachment 2.

VI. ENVIRONMENTAL COMMITMENTS AND MONITORING

The following environmental and monitoring commitments will be carried out under the preferred alternative or any of the other restricted fluctuating or steady flow alternatives described in the final EIS. A detailed description of these commitments can be found on pages 33 - 43 of that document. All practicable means to avoid or minimize environmental harm from the preferred alternative have been adopted.

1. Adaptive Management: This commitment includes the establishment of an Adaptive Management Workgroup, chartered in accordance with the Federal Advisory Committee Act; and development of a long-term monitoring, research, and experimental program which could result in some additional operational changes. However, any operational changes will be carried out in compliance with NEPA.

2. Monitoring and Protection of Cultural Resources: Cultural sites in Glen and Grand Canyons include prehistoric and historic sites and Native American traditional use and sacred sites. Some of these sites may erode in the future under any EIS alternative, including the no action alternative. Reclamation and the National Park Service, in consultation with Native American Tribes, will develop and implement a long-term monitoring program for these sites. Any necessary mitigation will be carried out according to a programmatic agreement written in compliance with the National Historic Preservation Act. This agreement is included as Attachment 5 in the final EIS.

3. Flood Frequency Reduction Measures: Under this commitment, the frequency of unanticipated floods in excess of 45,000 cubic feet per second will be reduced to an average of once in 100 years. This will be accomplished initially through the Annual Operating Plan process and eventually by raising the height of the spillway gates at Glen Canyon Dam 4.5 feet.

4. Beach/Habitat-Building Flows: Under certain conditions, steady flows in excess of a given alternative's maximum will be scheduled in the spring for periods ranging from 1 to 2 weeks. Scheduling, duration, and flow magnitude will be recommended by the Adaptive Management Work Group and scheduled through the Annual Operating Plan process. The objectives of these flows are to deposit sediment at high elevations, reform backwater channels, deposit nutrients, restore some of the natural system dynamics along the river corridor, and help the National Park Service manage riparian habitats.

5. New Population of Humpback Chub: In consultation with the U.S. Fish and Wildlife Service (FWS), National Park Service, and Arizona Game and Fish Department (AGFD), Reclamation will make every effort (through funding, facilitating, and technical support) to ensure that a new population of humpback chub is established in the mainstem or one or more of the tributaries within Grand Canyon.

6. Further Study of Selective Withdrawal: Reclamation will aggressively pursue and support research on the effects of multilevel intake structures at Glen Canyon Dam and use the results of this research to decide whether or not to pursue construction. FWS, in consultation with AGFD, will be responsible for recommending to Reclamation whether or not selective withdrawal should be implemented at Glen Canyon Dam. Reclamation will be responsible for design, NEPA compliance, permits, construction, operation, and maintenance.

7. Emergency Exception Criteria: Operating criteria have been established to allow the Western Area Power Administration to respond to various emergency situations in accordance with their obligations to the North American Electric Reliability Council. This commitment also provides for exceptions to a given alternative's operating criteria during search and rescue situations, special studies and monitoring, dam and power plant maintenance, and spinning reserves.

VII. BASIS FOR DECISION

The goal of selecting a preferred alternative was not to maximize benefits for the most resources, but rather to find an alternative dam operating plan that would permit recovery and long-term sustainability of downstream resources while limiting hydropower capability and flexibility only to the extent necessary to achieve recovery and long-term sustainability.

Based on the impact analysis described in the final EIS, three of the alternatives are considered to be environmentally preferable. They are: the Moderate Fluctuating Flow Alternative, the Modified Low Fluctuating

Flow Alternative, and the Seasonally Adjusted Steady Flow Alternative. Modified Low Fluctuating Plow is selected for implementation because it satisfies the critical needs for sediment resources and some of the habitat needs of native fish, benefits the remaining resources, and allows for titure hydropower flexibility, although there would be moderate to potentially major adverse impacts on power operations and possible decreases in long-term firm power marketing.   Nearly all-downstream resources are dependent to some extent on the sediment resource. This alternative meets the critical requirements of the sediment resource by restoring some of the pre-dam variability through floods and by providing a long-term balance between the supply of sand from Grand Canyon tributaries and the sand-transport capacity of the river. This, in turn, benefits the maintenance of habitat. The critical requirements for native fish are met by pursuing a strategy of warming releases from Glen Canyon Dam, enhancing the sediment resource, and substantially limiting the daily flow fluctuations.

The decision process for selecting the preferred alternative for the EIS followed a repetitive sequence of comparisons of effects on downstream resources resulting from each alternative.   Alternatives resulting in unacceptable adverse effects on resources (such as long-term loss of sandbars leading to the destruction of cultural resource sites and wildlife habitat) were eliminated from further comparisons. Comparisons continued until existing data were no longer available to support assumed benefits.

All resources were evaluated in terms of both positive and adverse effects from proposed alternatives.   Once it was determined that all alternatives would deliver at least 8.23 million acre feet of water annually, water supply played a minor role insubsequent resource evaluations. (One of the objectives of the "Criteria for Coordinated Long-Range Operation of Colorado River Reservoirs" is a minimum annual release of 8.23 million-acre feet of water from Glen Canyon Dam.) The alternatives covered a range of possible dam operations from maximum utilization of peaking power capabilities with large daily changes in downstream river levels (Maximum Power-plant Capacity Alternative) to the Year-Round Steady Flow Alternative that would have eliminated all river fluctuations and peaking power capabilities. Within this range, the Maximum Powerplant Capacity, No Action, and High Fluctuating Flow alternatives were eliminated from consideration as the preferred alternative because they would not meet the first criterion of resource recovery and long-term sustainability.   Data indicated that while beneficial to hydropower production, these alternatives would either increase or maintain conditions that resulted in adverse impacts to downstream resources under no action. For example, under these alternatives, the sediment resource would not likely be maintained over the long-term.

At the other end of the range, the Year-Round Steady Flow Alternative was also eliminated from consideration as the preferred alternative. This alternative would result in the greatest storage of sand within the river channel, the lowest elevation sandbars, the largest potential expansion of riparian vegetation, and the highest white-water boating safety benefits. However, it would not provide the variability on which the natural processes of the Grand Canyon are dependent (e.g. beach building, unvegetated sandbars, and backwater habitats). A completely stable flow regime would encourage the growth of vegetation thereby reducing bare-sand openings and patches of emergent marsh vegetation. This would limit beach camping and reduce the habitat value of these sites. With respect to other resources, this alternative did not provide any benefits beyond those already provided by other alternatives. Steady flows could also increase the interactions between native and non-native fish by intensifying competition and predation by non-natives on native fish. Such interactions would reach a level of concern under steady flows. Finally, this alternative would have major adverse impacts on hydropower (power operations and marketing).

The Existing Monthly Volume Steady Flow Alternative was eliminated from selection as the preferred alternative for reasons similar to those discussed above for the Year-Round Steady Flow Alternative.

Although the Interim Low Fluctuating Flow Alternative performed well over the interim period (August 1991 to the present), long-term implementation of this alternative would not restore some of the pre-dam variability in the natural system. The selected Modified Low Fluctuating Flow Alternative is an improved version of the Interim Low Fluctuating Flow Alternative because it would provide for some pre-dam variability through habitat maintenance flows.

The three remaining alternatives-the Moderate Fluctuating, Modified Low Fluctuating, and Seasonally Adjusted Steady Flow Alternatives- provide similar benefits to most downstream resources (e.g.. vegetation, terrestrial wildlife, and cultural resources) with respect to increased protection or improvement of those resources (see Table II-7 in the EIS). The Moderate Fluctuating Flow Alternative provided only minor benefits to native fish over no action conditions because of the relative similarity in flow fluctuations; and the benefits from the Seasonally Adjusted Steady Flow Alternative were uncertain given the improvement in habitat conditions for non-native fish this alternative would provide. Seasonally adjusted steady flows also would create conditions significantly different from those under which the current aquatic ecosystem has developed in the last 30 years and would adversely affect hydropower to a greater extent than the other two alternatives. The Modified Low Fluctuating Flow could substantially improve the aquatic food base and benefit native and non-native fish. The potential exists for a minor increase in the native fish population.

Although the Moderate Fluctuating, Modified Low Fluctuating, and Seasonally Adjusted Steady Flow Alternatives provide similar benefits to most downstream resources, the Modified Low Fluctuating Flow Alternative was selected as the preferred alternative because it would provide the most benefits with respect to the original selection criteria, given existing information. This alternative would create conditions that promote the protection and improvement of downstream resources while maintaining some flexibility in hydropower production. Although there would be a significant loss of hydropower benefits due to the selection of the preferred alternative (between $15.1 and $44.2 million annually) a recently completed non-use value study conducted under the Glen Canyon Environmental Studies indicates that the American people are willing to pay much more than this loss to maintain a healthy ecosystem in the Grand Canyon. The results of this non-use value study are summarized in Attachment 3 of the ROD.

The results of a General Accounting Office (GAO) audit mandated by the Grand Canyon Protection Act are in Attachment 4 of the ROD. This audit generally concludes that Reclamation used appropriate methodologies and the best available information in determining the potential impact of various dam flow alternatives on important resources. However, GAO identified some shortcomings in the application of certain methodologies and data, particularly with respect to the hydropower analysis. Reclamation's assumptions do not explicitly include the mitigating effect of higher electricity prices on electricity demand (price elasticity). GAO also determined that Reclamation's assumptions about natural gas prices were relatively high and that two computational errors were made during the third phase of the power analysis. According to GAO, these limitations suggest that the estimated economic impacts for power are subject to uncertainty. GAO also found limitations with some of the data used for impact analysis, Certain data was incomplete or outdated, particularly data used in assessing the economic impact of alternative flows on recreational activities. Nevertheless, the National Research

Council peer reviewed both the Glen Canyon Environmental Studies and the EIS, and generally found the analysis to be adequate. The GAO audit concluded that these shortcomings and limitations are not significant and would not likely alter the findings with respect to the preferred alternative and usefulness of the document in the decision-making process. The audit also determined that most of the key parties (83 percent of respondents) support Reclamation's preferred alternative for dam operations, although some concerns remain.

ATTACHMENT 1.

Specific peer reviewed sediment studies:

Beus, S. and C. Avery 1993. The influence of variable discharge regimes on Colorado River sand bars below Glen Canyon Dam. Glen Canyon Environmental Studies, Report PHYO101, Chapters I through 7 Northern Arizona University, Flagstaff, Ariz.

Beus, S., M.A. Kaplinski, J.E. Hazel, L. A. Tedrow, and L. H. Kearsley. 1995. Monitoring the effects of interim flows from Glen Canyon Dam on sand bar dynamics and campsite size in the Colorado River corridor, Grand Canyon National Park, AZ. Glen Canyon Environmental Studies, Report PHY 0112. Northern Arizona University, Flagstaff, Ariz.

Budhu, M and R. Gobin. 1994. Monitoring of sand bar instability during the interim flows: a seepage erosion approach. Glen Canyon Environmental Studies, Report PHY 0400. University of Arizona, Tucson, Ariz.

Carpenter, M., R. Carruth, Fink, D. Boling, and B. Cluer. 1995. Hydrogeology of sand bars 43.1 and 172.3L and the implications on flow alternatives along the Colorado River in the Grand Canyon. Glen Canyon Environmental Studies, Report PI-IY 0805. U.S. Geological Survey, Tucson, Ariz.

Cluer, B. 1993. Annual Report. Sediment mobility within eddies and the relationship to rapid erosion events. Glen Canyon Environmental Studies, Report PHY 0 11. National Park Service, Ft. Collins, Colo.

Cluer, B. and L. Dexter. 1994. An evaluation of the effects of the interim flows from Glen Canyon Dam on the daily change of beach area in Grand Canyon, Arizona. Glen Canyon Environmental Studies, Report PHY 0109. Northern Arizona University, Flagstaff, Ariz.

Nelson, J., N. Andrews, and J. MacDonald. 1993. Movement and deposition of sediments from the main channel to the eddies of the Colorado River in the Grand Canyon. Glen Canyon Environmental Studies, Report PI-W 0800. U.S. Geological Survey, Boulder, Colo.

Randle, T. J., RI. Strand, and A. Streifel. 1993. Engineering and environmental considerations of Grand Canyon sediment management. In: Engineering Solutions to Environmental Challenges: Thirteenth Annual USCOLD Lecture, Chattanooga, Tenn. U.S. Committee on Large Dams, Denver, Colo.

Schmidt, J. 1994. Development of a monitoring program of sediment storage changes in alluvial banks and bars, Colorado River, Grand Canyon, Arizona. Glen Canyon Environmental Studies, Report PI-W 0401. Utah State University. *

Smith, J. and S. Wiele. 1994. Draft report. A one-dimensional unsteady model of discharge waves in the Colorado River through the Grand Canyon. Glen Canyon Environmental Studies, Report PHY 0805. U.S. Geological Survey, Boulder, Colo.

Werrell, W., R. Ingliss, and L. Martin. 1993. Beach face erosion in Grand Canyon National Park: A response to ground water seepage during fluctuating flow releases from Glen Canyon Dam. Glen Canyon Environmental Studies, Report PHI' 0101, Chapter 4 k The influence of variable discharge regimes on Colorado River sandbars below Glen Canyon Dam, Report PHY 0 101. National Park Service, Ft. Collins, Colo.

Appendix E

Table II-7 Summary Comparison of Alternatives and Impacts

Table 11-7. – Summary Comparison of Alternatives and Impacts

	No Action	Maximum Powerplant Capacity	High Fluctuating Flow	Moderate Fluctuating Flow
WATER				
Streamflows (1,000 acre-feet)				
Annual streamflows				
Median annual release	8,753	8,573	8,559	8,559
Monthly streamflows (median)				
Fall (October)	568	568	568	568
Winter (January)	899	899	899	899
Spring (May)	587	587	592	592
Summer (July)	1,045	1,045	1,045	1,045

Hourly streamflows can be found in table 11-2.

	No Action	Maximum Powerplant Capacity	High Fluctuating Flow	Moderate Fluctuating Flow
SEDIMENT				
Riverbed sand (percent probability of net gain)				
After 20 years	50	49	53	61
After 50 years	41	36	45	70
Sandbars (feet)				
Active width	44 to 74	47 to 77	33 to 53	28 to 47
With habitat maintenance flows				41 to 66
Potential height	10 to 15	10 to 16	7 to 11	6 to 10
With habitat maintenance flows				9 to 14

	No Action	Maximum Powerplant Capacity	High Fluctuating Flow	Moderate Fluctuating Flow
FISH				
Aquatic food base	Limited by reliable wetted perimeter	Same as no action	Minor increase	Moderate increase
Native food	Stable to declining	Same as no action	Same as no action	Same as no action
Non-native warmwater and coolwater fish	Stable to declining	Same as no action	Same as no action	Same as no action
Interactions between native and non-native fish	Some predation and competition by non-natives	Same as no action	Same as no action	Same as no action

Trout	No Action	Maximum Powerplant Capacity	High Fluctuating Flow	Moderate Fluctuating Flow	
	Stocking-dependent	Same as no action	Same as no action	Increased growth potential, stocking-dependent	
	Modified Low Fluctuating Flow	Interim Low Fluctuating Flow	Existing Monthly Volume Steady Flow	Seasonally Adjusted Steady Flow	Year-Round Steady Flow
WATER					
Streamflows (1,000 acre-feet)					
Annual streamflows					
Median and annual release	8,559	8,559	8,559	8,554	8,578
Monthly streamflows (median)					
Fall (October)	568	568	568	492	699
Winter (January)	899	899	899	688	703
Spring (May)	592	592	592	1,106	699
Summer (July)	1,045	1,045	1,045	768	699

Hourly streamflows can be found in table 11-2.

	Modified Low Fluctuating Flow	Interim Low Fluctuating Flow	Existing Monthly Volume Steady Flow	Seasonally Adjusted Steady Flow	Year-Round Steady Flow
SEDIMENT					
Riverbed sand (percent probability of net gain)					
After 20 years	64	69	71	71	74
After 50 years	73	76	82	82	100
Sandbars (feet)					
Active width	24 to 41	24 to 41	10 to 19	16 to 29	0
With habitat maintenance flows	41 to 66			37 to 60	
Potential height	6 to 9	6 to 9	3 to 5	4 to 7	0 to 1
With habitat maintenance flows	9 to 14			8 to 13	
FISH					
Aquatic Fish Base	Potential major increase	Potential major increase	Major increase	Major increase	Major increase

	Modified Low Fluctuating Flow	Interim Low Fluctuating Low	Existing Monthly Volume Steady Flow	Seasonally Adjusted Steady Flow	Year-Round Steady Flow
Native Food	Potential minor increase	Potential minor increase	Uncertain potential minor increase	Uncertain potential minor increase	Uncertain potential minor increase
Non-native warmwater and coolwater fish	Potential minor increase	Potential minor increase	Potential minor increase	Potential minor increase	Potential minor increase
Interactions between native and non-native fish	Potential minor increase in warm, stable microhabitats	Potential minor increase in warm, stable microhabitats	Potential minor increase in warm, stable microhabitats	Potential minor increase in warm, stable microhabitats	Potential minor increase in warm, stable microhabitats
Trout	Increased growth potential, stocking-dependent	Increased growth potential, stocking-dependent	Increased growth potential, possibly self-sustaining	Increased growth potential, possibly self-sustaining	Increased growth potential, possibly self-sustaining

Table 11-7. – Summary Comparison of Alternatives and Impacts–Continued

	No Action	Maximum Powerplant Capacity	High Fluctuating Flow	Moderate Fluctuating Flow
VEGETATION				
Woody plants (area)				
New high water zone	No net change	0 to 9% reduction	15 to 35% increase	23 to 40% increase
With habitat maintenance flows				0 to 12% increase
Species composition	Tamarisk and others dominate	Tamarisk and others dominate	Tamarisk, coyote willow, arrowweed, and camelthorn dominate	Tamarisk, coyote willow, arrowweed, and camelthorn dominate
Emergent marsh plants				
New high water zone	No net change	Same as no action	Same as or less than no action	Same as or less than no action
Aggregate area of wet marsh plants				
WILDLIFE AND HABITAT				
Riparian habitat	*See vegetation.*			
Wintering waterfowl (aquatic food base)	Stable	Same as no action	Same as no action	Potential increase
ENDANGERED AND OTHER SPECIAL STATUS SPECIES				
Humpback chub	Stable to declining	Same as no action	Same as no action	Same as no action
Razorback sucker	Stable to declining	Same as no action	Same as no action	Same as no action
Flannelmouth sucker	Stable to declining	Same as no action	Same as no action	Same as no action
Bald eagle	Stable	Same as no action	Same as no action	Potential increase
Peregrine falcon	No effect	No effect	No effect	No effect
Kanab ambersnail	No effect	Some incidental take	Some incidental take	Some incidental take
Southwestern willow flycatcher	Undetermined increase	Same as no action	Same as no action	Same as no action

	Modified Low Fluctuating Flow	Interim Low Fluctuating Low	Existing Monthly Volume Steady Flow	Seasonally Adjusted Steady Flow	Year-Round Steady Flow
VEGETATION					
Woody plants (area)					
New high water zone	30 to 47% increase	30 to 47% increase	45 to 65% increase	38 to 58% increase	63 to 94% increase
With habitat maintenance flows	0 to 12% increase			0 to 12% increase	
Species composition	Tamarisk, coyote willow, arrowweed, and camelthorn dominate	Tamarisk, coyote willow, arrowweed, and camelthorn dominate	Tamarisk, coyote willow, arrowweed, and camelthorn dominate	Tamarisk, coyote willow, arrowweed, and camelthorn dominate	Tamarisk, coyote willow, arrowweed, and camelthorn dominate
Emergent marsh plants					
New high water zone					
Aggregate area of wet marsh plants	Same as or less than no action	Same as or less than no action	Less than no action	Less than no action	Less than no action
WILDLIFE AND HABITAT					
Riparian habitat	*See vegetation.*				
Wintering waterfowl (aquatic food base)	Potential increase	Potential increase	Potential increase	Potential increase	Potential increase
ENDANGERED AND OTHER SPECIAL SPECIES					
Humpback chub	Potential minor increase	Potential minor increase	Uncertain potential minor increase	Uncertain potential major increase	Uncertain potential minor increase
Razorback sucker	Potential minor increase	Potential minor increase	Uncertain potential minor increase	Uncertain potential minor increase	Uncertain potential minor increase
Flannelmouth sucker	Potential minor increase	Potential minor increase	Uncertain potential minor increase	Uncertain potential major increase	Uncertain potential minor increase
Bald eagle	Potential increase	Potential increase	Potential increase	Potential increase	Potential increase
Peregrine falcon	No effect	No effect	No effect	No effect	No effect
Kanab ambersnail	Some incidental take	Some incidental take	Some incidental take	Some incidental take	Some incidental take
Southwestern willow flycatcher	Same as no action	Same as no action	Same as no action	Same as no action	Same as no action

Table 11-7. – Summary Comparison of Alternatives and Impacts–Continued

	No Action	Maximum Powerplant Capacity	High Fluctuating Flow	Moderate Fluctuating Flow
CULTURAL RESOURCES				
Archeological sites (Number affected)	Major (336)	Major (336)	Potential to become major (263)	Moderate (Less than 157)
Traditional cultural properties	Major	Same as no action	Potential to become major	Moderate
Traditional cultural resources	Major	Same as no action	Same as no action	Increased protection
AIR QUALITY				
Regional air quality				
Total emissions (thousand tons)				
Sulfur dioxide	1,960	Same as no action	Slight reduction	Slight reduction
Nitrogen oxides	1,954			
RECREATION				
Fishing				
Angler activity	Potential danger	Same as no action	Same as no action	Moderate improvement
Day rafting Navigation past 3-Mile Bar	Difficult at low flows	Same as no action	Negligible improvement	Major improvement
White-water boating Safety	High risk at very high and very low flows	Same as no action	Negligible improvement	Minor improvement
Camping beaches (average area at normal peak stage)	Less than 7,720 square feet	Same as no action	Same as no action	Minor increase
Wilderness values	Influenced by range of daily fluctuations	Same as no action	Minor increase	Moderate increase
Economic benefits Change in equivalent annual net benefits (1991 nominal $ million)	0	0	0	+0.4

	No Action	Maximum Powerplant Capacity	High Fluctuating Flow	Moderate Fluctuating Flow	
Present value (1991 $ million)	0	0	0	+4.6	
	Modified Low Fluctuating Flow	Interim Low Fluctuating Low	Existing Monthly Volume Steady Flow	Seasonally Adjusted Steady Flow	Year-Round Steady Flow
CULTURAL RESOURCES					
Archeological sites (Number affected)	Moderate (Less than 157)	Moderate (Less than 157)	Moderate (Less than 157)	Moderate (Less than 157)	Moderate (Less than 157)
Traditional cultural properties	Moderate	Moderate	Moderate	Moderate	Moderate
Traditional cultural resources	Increased protection	Increased protection	Increased protection	Increased protection	Increased protection
WILDLIFE AND HABITAT					
Regional air quality					
Total emissions (thousand tons)					
Sulfur dioxide	Slight reduction	Slight reduction	Slight reduction	Slight reduction	Slight reduction
Nitrogen oxides					
RECREATION					
Fishing					
Angler safety	Moderate improvement	Moderate improvement	Major improvement	Major improvement	Major improvement
Day rafting					
Navigation past 3-mile bar	Major improvement	Major improvement	Major improvement	Major improvement	Major improvement

	Modified Low Fluctuating Flow	Interim Low Fluctuating Low	Existing Monthly Volume Steady Flow	Seasonally Adjusted Steady Flow	Year-Round Steady Flow
White-water boating Safety	Minor improvement	Minor improvement	Moderate improvement	Potential to become major increase	Major improvement
Camping beaches (average area at normal peak stage)	Minor increase	Minor increase	Major increase	Potential to become major increase	Major increase
Wilderness values	Moderate to potential to become major increase	Moderate to potential to become major increase	Major increase	Major increase	Major increase
Economic benefits Change in equivalent annual net benefits (1991 nominal $ million)	+3.7	+3.9	+3.9	+4.8	+2.9
Present value (1991 $ nominal million)	+43.3	+45.6	+45.6	+55.0	+23.5

Table 11-7. – Summary Comparison of Alternatives and Impacts–Continued

	No Action	Maximum Powerplant Capacity	High Fluctuating Flow	Moderate Fluctuating Flow
POWER				
Annual economic cost 1991 nominal $ million				
Hydrology	0	-1.5	2.1	54.0
Contract rate of delivery	0	0	2.5	36.7
Present value (1991 $ million)				
Hydrology	0	-17.3	24.3	624.5
Contract rate of delivery	0	0	28.9	424.5
Wholesale rate (1991 mills/kWh)	18.78	18.78	19.38 (+3.2%)	22.82 (+21.5%)
Retail rate (1991 mills/kWh)				
70% of end users	No change	No change	No change to slight decrease	No change to slight decrease
23% of end users	No change	No change	Slight decrease to moderate increase	Slight decrease to moderate increase
7% of end users (weighted mean)	64.1	64.1	64.6 (+0.8%)	69.7 (+8.8%)
NON-USE VALUE	*No data.*			

	Modified Low Fluctuating Flow	Interim Low Fluctuating Low	Existing Monthly Volume Steady Flow	Seasonally Adjusted Steady Flow	Year-Round Steady Flow
POWER					
Annual economic cost 1991 nominal $ million					
Hydrology	15.1	36.3	65.9	88.3	69.7
Contract rate of delivery	44.2	35.6	68.7	123.5	85.7

	Modified Low Fluctuating Flow	Interim Low Fluctuating Low	Existing Monthly Volume Steady Flow	Seasonally Adjusted Steady Flow	Year-Round Steady Flow
Present value (1991 $ million)					
Hydrology	174.6	418.7	761.4	1,021.2	805.0
Contract rate of delivery	511.2	411.7	794.6	1,428.4	991.2
Wholesale rate (1991 mills/kWh)	23.16 (+23.3%)	23.18 (23.4%)	25.22 (34.4%)	28.20 (+50.2%)	26.78 (+42.6%)
Retail rate (1991 mills/kWh)					
70% of end users	No change to slight decrease	No change to slight decrease	No change to slight decrease	No change to slight decrease	No change to slight decrease
23% of end users	Slight decrease to moderate increase	Slight decrease to moderate increase	Slight decrease to moderate increase	Slight decrease to moderate increase	Slight decrease to moderate increase
7% of end users (weighted mean)	70.5 (+10.0%)	70.2 (+9.6%)	72.9 (+13.8%)	75.8 (+18.4%)	74.5 (+16.3%)
NON-USE VALUE	*No data.*				

Appendix F

Economic Literature Relevant to Grand Canyon Management

An inventory of economic valuation literature about which Grand Canyon managers should be aware is provided in this appendix. Only recently published papers are considered. A longer history of economic research on any particular topic can be obtained by searching online databases, such as the EconLit database (a database of all articles published in economics journals since 1969, although the very early articles do not include abstracts) provided by the American Economic Association.

The topics covered include studies specific to the Grand Canyon, studies of the Columbia and other river systems, and studies of the valuation of relevant environmental goods in other contexts (hydropower, ecosystems, national park recreation, guided rafting, recreational fishing, hiking, waterfowl hunting, biodiversity and endangered species, cultural heritage, water quality, streamflow, and geographic scope). Not all these papers have yet been retrieved and formally evaluated. In some cases, their abstracts are relied upon as a guide to their content. Also included are papers on relevant big-picture issues, including the evolving debate about nonmarket environmental valuation, methodologies for valuation, and the ethical and philosophical issues involved. An accessible overview of the principles of economic benefit–cost analysis in the context of water resources is likely provided by Griffin (1998).

Studies Specific to the Grand Canyon

It is soon apparent in any search of the literature that very little of the economics research on environmental values published in the last few years has focussed on Grand Canyon resources directly relevant to the strategic planning process. The closest thing appears to be a study by Champ et al. (1997) which explores real and hypothetical donations scenarios in a survey used to assess the nonuser social value of a program at Grand Canyon National Park to remove compacted dirt roads on the North Rim of the Canyon.

Studies of the Columbia and Other River Systems

Issues of water resource management on the Columbia River system have generated a fair amount of research in environmental economics. For example, McGinnis (1995) discusses the economic conflicts between wild salmon and power generation, and a Washington State University Ph.D. dissertation by Reilly (1995) considers multiple-criteria decision-making in the context of the Columbia River Basin Salmon Recovery Plan.

Anderson et al. (1993) present a multidisciplinary study that focuses on the influence of climate change, but explores the effects of habitat changes on the production and economic value of spring chinook salmon in the Yakima River tributary of the Columbia in eastern Washington. The total economic value of a fish is the sum of its existence, commercial, recreational, and capital values, and the change in total economic value per fish associated with reducing one fish run is found to be significant.

Burtraw and Frederick (1993) consider how compensation principles might be used to promote political support, in ways consistent with equity and efficiency goals, for the Idaho drawdown plan, proposed to protect and restore Snake River salmon. Notably, it does not evaluate the efficacy or cost-effectiveness of the plan.

Cameron et al. (1996) provide an empirical model of recreation demand on federal reservoirs and run-of-river projects in the Columbia River basin as a function of water levels or flow rates, estimated using actual and intended participation data under real and counterfactual conditions.

For river systems other than the Columbia, there may be useful insights in Cordell and Bergstrom (1993), which examines the social value of recreational water uses under alternative reservoir-level-management scenarios. Alternatively, Garrod and Willis (1996) study the social value of environmental enhancement on the River Darent.

Other recent research is reported in a University of Tennessee study by Murray et al. (1998), concerning the economic effects of drawdowns on the Cherokee and Douglas lakes in the southeast United States.

Studies of Relevant Issues in Other Contexts

Hydropower

The problem of the humpback chub in the Grand Canyon may have some features in common with the management of salmon populations in the Columbia River system, although a key difference is the commercial value of the two species. Paulsen and Wernstedt (1995) describe simulation and optimization models designed to analyze the cost-effectiveness of salmon recovery measures required by the 1980 Northwest Power Planning Act. The competition between ecological viability and "economic" uses of aquatic systems is examined by Teclaff and Teclaff (1994), which looks at the problem of ecosystems that have been and are being damaged by waterworks projects constructed for "economically" beneficial purposes. This study considers examples of damage to ecosystems in the past, what restoration techniques are currently being used, and relevant developments in domestic and international water law and policy.

Ecosystems

Wetlands are far from the most significant issue in the allocation of Grand Canyon water resources, but wetlands debates in other contexts have spawned a considerable amount of research concerning how to quantify the social values of a complex set of ecological functions represented by one type of water-based resource. For example, Bateman and Langford (1997) examine nonusers' values for preserving the Norfolk Broads, a wetland area in the United Kingdom of recognized international importance, from the threat of saline flooding. Earlier, Cooper and Loomis (1993) examined whether water deliveries to wetlands have a systematic effect on the level of waterfowl hunting benefits. Additional research on wetlands valuation is currently underway at the University of Iowa.

National Park Recreational Values

The fact that the Grand Canyon attracts a very significant number of recreational users means that studies of the recreational values associated with national parks, both in the United States and abroad, are also relevant.

Beal (1995) describes an application of the travel cost method to assess the recreational value for both camping and day visits to Carnarvon Gorge National Park. Kosz (1996) finds that only 20 percent of the willingness to pay measured by contingent valuation is needed to make the net present value of the "best" national park policy variant in an Austrian study equal to that of the "best" hydroelectric power policy variant. Chase et al. (1998) estimate demand for ecotourism in national parks in Costa Rica.

Guided Rafting Values

Because white-water rafting is such a significant component of total recreational use, it is also important to follow studies that characterize demand for this type of activity in contexts other than the Grand Canyon. Travel cost methods are used by Bowker et al. (1996) to estimate the nonmarket economic "user" value of guided white-water rafting on two southern rivers. For Ontario's wilderness parks, Rollins (1997) uses contingent valuation methods to assess user benefits from wilderness canoeing, an experience that has some features in common with some of the recreational uses of the Grand Canyon.

Shaw and Jakus (1996) make an important contribution to the recreational nonmarket valuation literature by explicitly considering the interaction between participant skill levels and resource attributes in the measurement of recreational values of nonmarket resources. Their application concerns technical rock-climbing, but the grading for technical difficulty for this recreational use is very similar to that for white-water rafting.

Recreational Fishing

Again, salmon are different from humpback chub, or even from trout, but a paper by Layman et al. (1996) can provide insights on how fisheries management strategies affect a recreational fishery. In this study, travel cost methods are used to examine how fisheries management tools affect the economic value of the recreational chinook salmon fishery in the Gulkana River of Alaska. There are bag limits and similar restrictions that might apply to the management of the Grand Canyon trout fishery and its consequent value to recreational fishers.

Grand Canyon management decisions also depend indirectly on how conditions in the Grand Canyon affect the value of the recreational fishery. While temperature may vastly dominate toxic contamination as a water-quality issue in the river below the Glen Canyon Dam, the way researchers have incorporated toxic contamination will be analogous to how temperature could be incorporated into a valuation model. In Montgomery and Needelman (1997), a repeated discrete choice model of fishing behavior was used to evaluate the welfare costs, to users, of toxic contamination in freshwater fish.

Hiking

Hiking may be a relatively smaller component of total recreational use of the Grand Canyon, but the work of Casey et al. (1995) will be relevant to the extent that hiking values figure in the social value of the Grand Canyon. The authors combine a standard travel cost survey design with a contingent valuation type question about willingness to accept compensation to forego access to a resource. Their work highlights the importance of correctly measuring the opportunity cost of time. In other work on the social value associated with hiking, the willingness to pay of Washington State residents for hiking opportunities in the Cascade Mountain Range was estimated by Englin and Shonkwiler (1995).

Waterfowl Hunting

Waterfowl hunting can be an important source of social value for water-based environmental resources in many contexts. Gan and Luzar (1993) describe the use of conjoint analysis, a generalization of contingent valuation, to analyze waterfowl hunting in Lousiana.

Biodiversity/Endangered Species

Concerns about endangered species (and/or the preservation of biodiversity) are clearly one of the key issues in Grand Canyon management. Several recent empirical studies will have some bearing on how policy-makers think about inferring social values for species such as the humpback chub. The late 1990s have seen a profusion of economic studies on these two related issues, presented alphabetically by author below:

- In Garrod and Willis (1997), in the context of forestry, not a riverine system, the potential nonuse value of programs to enhance biodiversity is assessed.
- Gowdy (1997) discusses the value of biodiversity at different levels, including market value, nonmarket values to humans, and the value of biodiversity to ecosystems, emphasizing the need for hierarchical and pluralistic methodology to determine appropriate policies for the preservation of biodiversity.
- Hanley et al. (1995) enumerate a variety of problems that researchers can generally expect to encounter in valuing the protection of biodiversity.
- Jakobsson and Dragun (1996) report on a comprehensive study for Victoria, Australia, of the social values associated with conservation of endangered native flora and fauna in general (and on the preservation of Leadbeater's possum in particular).
- Loomis and White (1996) describe a systematic review (metaanalysis) of a variety of estimates of the economic benefits of rare and endangered species.
- Loomis and Ekstrand (1997) describe a contingent valuation study to estimate the economic benefits of preserving critical habitat for the Mexican Spotted Owl. Ekstrand and Loomis (1998) explore the role

of respondent uncertainty in the estimation of economic benefits of protecting critical habitat for nine threatened and endangered fish species living in the Colorado, Green, and Rio Grande River basins.

- Macmillan et al. (1996) describe the use of contingent valuation to assess willingness to pay to prevent uncertain biodiversity losses in upland areas of Scotland important for nature conservation.

- Metrick and Weitzman (1996) present a statistical analysis of the main determinants of government decisions about the preservation of endangered species, finding that the role of "visceral" characteristics (size, being a "higher form of life") dominate "scientific" characteristics (degree of endangerment, taxonomic uniqueness).

- Perrings et al. (1995) assemble ten papers that consider what is at issue in the problem of biodiversity loss, with most authors being economists.

- Polasky and Solow (1995) describe models they developed for the valuation of a collection of potentially beneficial species; the models acknowledge that beneficial species are not perfect substitutes and that the probability that each species is beneficial may depend on the outcome for other species.

- Shogren and Crocker (1995) review the challenges of valuing ecosystems and biodiversity in a volume devoted to Great Plains ecosystems.

- Simon and Doerksen (1995) consider the implications of differential spending on endangered species recovery and the priority ranking assigned by the U.S. Fish and Wildlife Service to particular species. A species recovery priority rank is not related to funding decisions, although some of its components (recovery potential and conflict with development) are correlated with funding. Funding is also greater for mammals, birds, and fish.

- Willis et al. (1996) report upon a case study of the Pevensey Levels in the United Kingdom in evaluating the benefits and costs of a wildlife enhancement scheme.

Cultural heritage; Native rights

The economics literature does not generally stray into the area of cultural values of resources. However, one exception is a study by Lockwood et al. (1996), which describes the use of contingent valuation methods to assess the trade offs between use of the Australian Alps for cattle grazing or for cultural heritage conservation. More recently, Adamowicz et al. (1998) explored the advisability of aggregating across indigenous and nonindigenous values in assessing the social values of environmental goods.

Lin et al. (1996) explore the welfare effects on recreational anglers of alternative salmon allocation policies to meet Native American treaty rights, using the Willamette River as an example.

Water Quality

Freeman (1995) reviews the empirical literature on the economic value of marine recreation fishing, beach visits, and boating. Considerable heterogeneity in effects of water quality is identified, and the links between pollution and key factors such as catch rates have not been sufficiently firmly established.

Streamflow

Loomis (1996) describes measurement of the economic benefits of removing dams and restoring the Elwha River, and Naeser and Smith (1995) consider the various issues involved in conflicts over in-stream flows on the Upper Arkansas River of Colorado, when environmental quality and recreational activity depend upon both the level and duration of river flows. Maintenance of in-stream flows is complicated by the nature of the water appropriations system. Harpman et al. (1993) also discuss the valuation of flow changes on a fishery.

More recently, for New Mexico, Berrens et al. (1998) describe the use of survey data to demonstrate strong evidence of public support for in-stream flow protection and its associated nonmarket benefits. This paper builds on other work described in Berrens et al. (1996). Loomis (1998) uses survey methods to measure the perceived benefits of in-stream flows for recreation and endangered fish. The value of stream-flow is also considered in Loomis (1997). Further, Douglas and Taylor (1998) examine in-stream flow benefits associated with the Trinity River using both indirect market and nonmarket valuation methods. Willis and Whittlesey (1998) present an integrated economic and hydrological model for measuring both the economic cost and hydrologic consequences of maintaining a minimum stream-flow level.

For Puerto Rico, Gonzalez Caban and Loomis (1997) describe a study of willingness to pay for preserving in-stream flows in the Rio Mameyes and avoiding a dam on the Rio Fajardo. The authors also addressed some methodological issues in valuation in Loomis and Gonzalez Caban (1997).

Geographic Scope

In Pate and Loomis (1997), willingness to pay (for programs designed to reduce various environmental problems in the San Joaquin valley in California) is explicitly modeled as a function of geographic distance from the affected resources. The results have implications for underestimation of benefits if the geographic extent of the public good in question is limited to one political jurisdiction.

Distributional Issues: Regional Economic Impacts

Traditional regional economic impact analysis with respect to river management, which is still being conducted, remains relevant to considerations of distributional effects of management decisions. A study by Leones et al. (1997) looks at the local effects of upstream diversions on the Rio Grande and the consequences for popular white-water runs, focusing on visitation and total expenditures but ignoring nonmarket benefits.

Water Allocation and Water Marketing

Gaffney (1997) addresses the issue of efficient allocation of water resources among competing end uses and explains why a market for raw water is necessary, yet why existing markets work badly. Optimal water-allocation issues with an application to the Nestos River in the Balkans are considered by Giannias (1997). A case study of Colorado River water allocations is employed by Mendelsohn and Bennett (1997) in their study of global warming and its potential consequences for water allocations.

The Evolving Debate about Nonmarket Environmental Valuation

The task of assessing credible nonmarket values for environmental goods remains controversial. A number of discussions and examples can provide a good sense of the debate and current progress towards its resolution. The debate "ramped up" dramatically as a result of the litigation surrounding the Exxon Valdez oil spill in 1989. Carson et al. (1994) describe contingent valuation as it was used in this high-profile litigation. This study, sponsored by the state of Alaska, as part of its natural resource damage assessment following the spill, was designed to elicit society's willingness to pay to prevent another such oil spill. For a contrary view, the volume by Hausman (1993) represents a compilation of research funded by Exxon in opposition to the use of contingent valuation methods in the litigation over the Exxon Valdez oil spill.

In Kopp and Pease (1996), the authors, who are generally identified with plaintiffs in environmental litigation, review the debate about the use of contingent valuation methods to measure total nonmarket values of environmental goods, touching on the academic, legal, and political issues. This paper is a useful counterpoint to the opinions expressed in the edited volume by Hausman (1993). A research team that is generally perceived as favorable to defendants in natural resources damages litigation, Dunford et al. (1997), examine the economic and legal constraints that determine whose losses are included (and whose should *not* be included) in natural resources damage assessment.

The distinction between "user" values and "nonuser" values of environmental goods is also relevant to Grand Canyon management. The Hagler Bailly (1997) study attempted to measure nonuse values, in an effort that produced estimates that were heavily criticized by some stakeholders. Cummings and Harrison (1995) provide a critical review of some of the key issues concerning nonuse values. Another useful reference for current thinking on the subject of nonmarket valuation is a monograph by Smith (1996).

Methodology of Valuation

The Bishop et al. (1993) studies and the Hagler Bailly nonuse value study no longer represent the state of the art in valuation of nonmarket environmental goods. There has been considerable innovation in this research area over the last several years, and much more is known about how to undertake valuation studies than is reflected in these earlier projects.

One of the key developments in the last few years has been a shift toward the use of "conjoint analysis" or "choice experiments" or "stated-preference" techniques. These methods generalize the earlier generation of referendum contingent valuation techniques and can readily accommodate observed choices as well as stated behavior in the same format.

For example, Boxall et al. (1996) compare contingent valuation methods for nonmarket valuation with "choice experiment" value elicitation methods imported from the marketing research and transportation literatures. Although their application is to recreational moose hunting values, their work highlights the importance of substitutes in environmental valuation and draws attention to this generalization of contingent valuation methods as a more appropriate technique in some cases. Likewise, Hanley et al. (1998) report on a study of the economic value of the conservation benefits of environmentally sensitive areas in Scotland, using both contingent valuation and choice experiment methods.

The theme of combining stated choices with observed choices has also strengthened in non-conjoint-analysis studies. Huang et al. (1997), for example, evaluate the strategy of identifying the social values for environmental quality improvements by combining individuals' observed choices with their claims about how they would behave in hypothetical choice situations. The objective is the development of theoretically consistent welfare measurements of use and nonuse values.

A couple of recent papers address respondent information and experience as determinants of nonmarket values for environmental goods. Hutchinson et al. (1995) examine the problems of information provision and respondent knowledge, comprehension, and cognition in the use of contingent valuation methods to measure nonuse values. Cameron and Englin (1997b) illustrate the effects of acknowledging each respondent's level of experience with a nonmarket resource upon the expected value and dispersion in estimated willingness-to-pay values for an environmental good (in this case, trout fishers and the prevention of acid rain damage to high-altitude lakes in the Northeast United States). Using other data from the same survey, these authors also develop a theoretically based empirical model of user/nonuser status and the dependence of individual values on this status (Cameron and Englin, 1997a). Similar issues are addressed in Niklitschek and Leon (1996), which estimates the total (use and nonuse) value of a resource under a capacity constraint, introducing information on intended use as an integral part of the contingent valuation method that is employed. The authors assert that this combined approach allows use and nonuse values to be distinguished for a sample of users and nonusers.

On the issue of nonuse values, there have been at least two relevant papers in the recent literature: Lazo et al. (1997) explore the issue of potential double-counting across generations as a criticism of nonuse value being included in benefits estimates, showing that the criticism is not warranted, and McConnell (1997) explores the validity of motives for existence or passive-use value, in particular, altruism. The author constructs models of three types of altruism and examines the implications when benefit–cost analysis must be conducted for a population of heterogeneous altruists.

Stevens et al. (1994) assess the temporal stability of contingent valuation bids for wildlife existence (finding that these values are relatively stable) but also consider the different possible interpretations of the values that are elicited.

Philosophy of Valuation

Nonmarket valuation of environmental goods sometimes brings neo-classical microeconomics into direct confrontation with psychology and philosophy, and this area of research must continue to be informed by other disciplines in order for progress to be made on its fundamental issues. For example, Brown (1994) argues that the process of estimating nonuse values requires that researchers have expertise in a number of disciplines, and Nelson (1997) questions whether environmental economics is beginning to encroach on religion by contemplating existence values. In a similar vein, the concept of nonuse value from the perspective of environmental philosophy is considered by Mazzotta and Kline (1995), and Crowards (1997) explores some of the ethical and economic motivations surrounding nonuse values.

As opposed to economic considerations, ethics is sometimes argued to underlie the values that survey respondents claim to hold for environmental goods. In Blamey et al. (1995), some doubt is cast upon the interpretation of contingent valuation results as characterizations of consumer preferences; instead it is argued that they reflect ethical concerns and that respondents are acting as "citizens" rather than "consumers."

There has also been considerable soul-searching about the whole challenge of inferring values for environmental goods that can be compared to the costs that must be borne by society in order to preserve them. O'Neill (1997), for example, note that ethical reasoning of all types is anthropocentric and considers different types of reasoning and their implications for protection of the natural world. Booth (1994) provides a monograph that addresses the complex of values underlying the decline and preservation of old-growth forest in the Pacific Northwest, including aboriginal use and treatment and the value system brought by European settlers.

O'Riordan (1997) argues that valuation is more than just static willingness to pay and asserts that valuation through economic measures can be built upon by creating trusting and legitimizing procedures of stakeholder negotiation and mediation.

Recent work in Australia by Cameron (1997), applied to water quality (integrated catchment management), employs a blend of nonmarket valuation and other environmental valuation philosophies.

REFERENCES

Adamowicz, W., T. Beckley, D. H. MacDonald, L. Just, M. Luckert, E. Murray, and W. Phillips. 1998. In search of forest resource values of indigenous peoples: Are nonmarket valuation techniques applicable? Society and Natural Resources 11(1):51–66.

Anderson, D. M., S. A. Shankle, M. J. Scott, D. A. Neitzel, and J. C. Chatters. 1993. Valuing effects of climate change and fishery enhancement on chinook salmon. Contemporary Policy Issues 11(4):82–94.

Arrow, K., R. Solow, P. R. Portney, E. E. Leamer, R. Radner, and H. Schuman. 1993. Report of the NOAA Panel on Contingent Valuation. Federal Register 58(10):4601-4614.

Bateman, I. J., and I. H. Langford. 1997. Non-users' willingness to pay for a national park: An application and critique of the contingent valuation method. Regional Studies 31(6):571–582.

Beal, D. J. 1995. A travel cost analysis of the value of Carnarvon Gorge National Park for recreational use. Review of Marketing and Agri-cultural Economics 63(2):292–303.

Berrens, R. P., P. Ganderton, and C. L. Silva. 1996. Valuing the protection of minimum instream flows in New Mexico. Journal of Agricultural and Resource Economics 21(2):294–308.

Berrens, R. P., A. Bohara, H. Jenkins-Smith, C. L. Silva, P. Ganderton, and D. Brookshire. 1998. A joint investigation of public support and public values: Case of instream flows in New Mexico. Ecological Economics 27(2):189–203.

Bishop, R. C., K. J. Boyle, and M. P. Welsh. 1993. The role of question order and respondent experience in contingent-valuation studies. Journal of Environmental Economics and Management 25(1):S80-S99.

Blamey, R., M. Common, and J. Quiggin. 1995. Respondents to contingent valuation surveys: Consumers or citizens? Australian Journal of Agricultural Economics 39(3):263–288.

Booth, D. E. 1994. Valuing nature: The decline and preservation of old-growth forests. London: Rowman and Littlefield.

Bowker, J. M., D. B. K. English, and J. A. Donovan. 1996. Toward a value for guided rafting on southern rivers. Journal of Agricultural and Applied Economics 28(2):423–432.

Boxall, P. C., W. L. Adamowicz, J. Swait, M. Williams, and J. Louviere. 1996. A comparison of stated preference methods for environmental valuation. Ecological Economics 18(3): 243–253.

Brown, G. 1994. Estimating nonuse values requires interdisciplinary research. Pg. 11–26 in Economy, Environment, and Technology: A Socio-Economic Approach. Armonk, N.Y.: Sharpe.

Burtraw, D., and K. D. Frederick. 1993. Compensation Principles for the Idaho Drawdown Plan. Washington, D.C.: Resources for the Future.

Cameron, J. I. 1997. Applying socio-ecological economics: A case study of contingent valuation and integrated catchment management. Ecological Economics 23(2):155–165.

Cameron, T. A., and J. Englin. 1997a. Respondent experience and contingent valuation of environmental goods. Journal of Environmental Economics and Management 33(3):296–313.

Cameron, T. A., and J. Englin. 1997b. Welfare effects of changes in environmental quality under individual uncertainty about use. RAND Journal of Economics 28(0):45–70.

Cameron, T. A., W. D. Shaw, S. E. Ragland, J. M. Callaway, and S. Keefe. 1996. Using actual and contingent behavior data with differing levels of time aggregation to model recreation demand. Journal of Agricultural and Resource Economics 21(1):130–149.

Carson, R. T., R. C. Mitchell, W. M. Haneman, R. J. Kopp, S. Presser, and P. A. Ruud. 1994. Contingent Valuation and Lost Passive Use: Damages from the Exxon Valdez. Washington, D.C.: Resources for the Future.

Casey, J. F., T. Vukina, and L. E. Danielson. 1995. The economic value of hiking: Further considerations of opportunity cost of time in recreational demand models. Journal of Agricultural and Applied Economics 27(2):658–668.

Champ, P. A., R. C. Bishop, T. C. Brown, and D. W. McCollum. 1997. Using Donation Mechanisms to Value Nonuse Benefits from Public Goods. Journal of Environmental Economics and Management 33(2):151–162.

Chase, L. C., D. R. Lee, W. D. Schulze, and D. J. Anderson. 1998. Ecotourism demand and differential pricing of national park access in Costa Rica. Land Economics 74(4):466–482.

Cooper, J., and J. Loomis. 1993. Testing whether waterfowl hunting benefits increase with greater water deliveries to wetlands. Environmental and Resource Economics 3(6):545–561.

Cordell, H. K., and J. C. Bergstrom. 1993. Comparison of recreation use values among alternative reservoir water level management scenarios. Water Resources Research 29(2):247–258.

Crowards, T. 1997. Nonuse values and the environment: Economic and ethical motivations. Environmental Values 6(2):143–167.

Cummings, R. G., and G. W. Harrison. 1995. The measurement and decomposition of nonuse values: A critical review. Environmental and Resource Economics 5(3):225–247.

Douglas, A. J., and J. G. Taylor. 1998. Riverine based eco-tourism: Trinity River non-market benefits estimates. International Journal of Sustainable Development and World Ecology 5(2):136–148.

Dunford, R. W., F. R. Johnson, and E. S. West. 1997. Whose losses count in natural resource damages? Contemporary Economic Policy 15(4):77–87.

Ekstrand, E. R., and J. Loomis. 1998 Incorporating respondent uncertainty when estimating willingness to pay for protecting critical habitat for threatened and endangered fish. Water Resources Research 34(11):3149–3155.

Englin, J., and J. S. Shonkwiler. 1995. Estimating social welfare using count data models: An application to long-run recreation demand under conditions of endogenous stratification and truncation. Review of Economics and Statistics 77(1):104–112.

Freeman, A. M., III. 1995. The benefits of water quality improvements for marine recreation: A review of the empirical evidence. Marine Resource Economics 10(4):385–406.

Gaffney, M. 1997. What price water marketing? California's new frontier. American Journal of Economics and Sociology 56(4):475–520.

Gan, C., and E. J. Luzar. 1993. A conjoint analysis of waterfowl hunting in Louisiana. Journal of Agricultural and Applied Economics 25(2):36–45.

Garrod, G. D., and K. G. Willis. 1996. Estimating the benefits of environmental enhancement: A case study of the River Darent. Journal of Environmental Planning and Management 39(2):189–203.

Garrod, G. D., and K. G. Willis. 1997. The non-use benefits of enhancing forest biodiversity: A contingent ranking study. Ecological Economics 21(1):45–61.

Giannias, D. A. 1997. Marginal values of water quantity and quality variations. Quality and Quantity 31(4):337–346.

Gonzalez Caban, A., and J. Loomis. 1997. Economic benefits of maintaining ecological integrity of Rio Mameyes, in Puerto Rico. Ecological Economics 21(1):63–75.

Gowdy, J. M. 1997. The value of biodiversity: Markets, society, and ecosystems. Land Economics 73(1):25–41.

Griffin, R. C. 1998. The fundamental principles of cost-benefit analysis. Water Resources Research 34(8):2063–2071.

Hagler Bailly Consulting, Inc. 1997. Glen Canyon Dam, Colorado River Storage Project, Arizona. Final Nonuse Values Study Summary Report. Madison, Wisc.

Hanley, N., C. Spash, and L. Walker. 1995. Problems in valuing the benefits of biodiversity protection. Environmental and Resource Economics 5(3):249–272.

Hanley, N., D. MacMillan, R. E. Wright, C. Bullock, I. Simpson, D. Parsisson, and B. Crabtree. 1998. Contingent valuation versus choice experiments: Estimating the benefits of environmentally sensitive areas in Scotland. Journal of Agricultural Economics 49(1):1–15.

Harpman, D. A., E. W. Sparling, and T. J. Waddle. 1993. A methodology for quantifying and valuing the impacts of flow changes on a fishery. Water Resources Research 29(3):575–582.

Hausman, J. A., ed. 1993. Contingent valuation: A critical assessment. Vol. 220, Contributions to Economic Analysis. New York: Elsevier.

Huang, J. C., T. C. Haab, and J. C. Whitehead. 1997. Willingness to pay for quality improvements: Should revealed and stated preference data be combined? Journal of Environmental Economics and Management 34(3):240–255.

Hutchinson, W. G., S. M. Chilton, and J. Davis. 1995. Measuring non-use value of environmental goods using the contingent valuation method: Problems of information and cognition and the application of cognitive questionnaire design methods. Journal of Agricultural Economics 46(1):97–112.

Jakobsson, K. M., and A. K. Dragun. 1996. Contingent valuation and endangered species: Methodological issues and applications. Chelten-ham, United Kingdow: Edward Elgar.

Kopp, R. J., and K. A. Pease. 1996. Contingent Valuation: Economics, Law and Politics. Washington, D.C.: Resources for the Future.

Kosz, M. 1996. Valuing Riverside wetlands: The case of the Donau-Auen National Park. Ecological Economics 16(2):109–127.

Layman, R. C., J. R. Boyce, and K. R. Criddle. 1996. Economic valuation of the chinook salmon sport fishery of the Gulkana River, Alaska, under current and alternate management plans. Land Economics 72(1):113–128.

Lazo, J. K., G. H. McClelland, and W. D. Schulze. 1997. Economic theory and psychology of non-use values. Land Economics 73(3):358–371.

Leones, J., B. Colby, D. Cory, and L. Ryan. 1997. Measuring regional economic impacts of streamflow depletions. Water Resources Research 33(4):831–838.

Lin, P. C., R. M. Adams, and R. P. Berrens. 1996. Welfare effects of fishery policies: Native American treaty rights and recreational salmon fishing. Journal of Agricultural and Resource Economics

21(2):263–276.

Lockwood, M., P. Tracey, and N. Klomp. 1996. Analysing conflict between cultural heritage and nature conservation in the Australian Alps: A CVM approach. Journal of Environmental Planning and Management 39(3):357–370.

Loomis, J. 1996. Measuring the economic benefits of removing dams and restoring the Elwha River: Results of a contingent valuation survey. Water Resources Research 32(2):441–447.

Loomis, J. B. 1997. Panel estimators to combine revealed and stated preference dichotomous choice data. Journal of Agricultural and Resource Economics 22(2):233–245.

Loomis, J. B. 1998. Estimating the public's values for instream flow: Economic techniques and dollar values. Journal of the American Water Resources Association 34(5):1007–1014.

Loomis, J. B., and E. Ekstrand. 1997. Economic benefits of critical habitat for the Mexican spotted owl: A scope test using a multiple-bounded contingent valuation survey. Journal of Agricultural and Resource Economics 22(2):356–366.

Loomis, J. B., and A. Gonzalez Caban. 1997. How certain are visitors of their economic values of river recreation: An evaluation using repeated questioning and revealed preference. Water Resources Research 33(5):1187–1193.

Loomis, J. B., and D. S. White. 1996. Economic benefits of rare and endangered species: summary and meta-analysis. Ecological Economics 18(3):197–206.

Macmillan, D., N. Hanley, and S. Buckland. 1996. A contingent valuation study of uncertain environmental gains. Scottish Journal of Political Economy 43(5):519–533.

Mazzotta, M. J., and J. Kline. 1995. Environmental philosophy and the concept of nonuse value. Land Economics 71(2):244–249.

McConnell, K. E. 1997. Does altruism undermine existence value? Journal of Environmental Economics and Management 32(1):22–37.

McGinnis, M. V. 1995. On the verge of collapse: The Columbia River system, wild salmon and the Northwest Power Planning Council. Natural Resources Journal 35(1):63–92.

Mendelsohn, R., and L. L. Bennett. 1997. Global warming and water management: Water allocation and project evaluation. Climatic Change 37(1):271–290.

Metrick, A., and M. L. Weitzman. 1996. Patterns of behavior in

Montgomery, M., and M. Needelman. 1997. The welfare effects of toxic contamination in freshwater fish. Land Economics 73(2):211–223.

Murray, M. N., V. Cunningham, P. Dowell, P. Jakus, and S. Proca. 1998. The Economic and Fiscal Consequences of TVA's Draw Down of Cherokee and Douglas Lakes: Final Project Report. Knoxville, Tenn.: Center for Business and Economic Research.

Naeser, R. B., and M. G. Smith. 1995. Playing with borrowed water: Conflicts over instream flows on the Upper Arkansas River. Natural Resources Journal 35(1):93–110.

Nelson, R. H. 1997. Does existence value exist? Environmental economics encroaches on religion. Independent Review 1(4):499–521.

Niklitschek, M., and J. Leon. 1996. Combining intended demand and yes/no responses in the estimation of contingent valuation models. Journal of Environmental Economics and Management 31(3):387–402.

O' Neill, O. 1997. Environmental values, anthropocentrism and speciesism. Environmental Values 6(2):127–142.

O' Riordan, T. 1997. Valuation as revelation and reconciliation. Environmental Values 6(2):169–183.

Pate, J., and J. Loomis. 1997. The effect of distance on willingness to pay values: A case study of wetlands and salmon in California. Ecological Economics 20(3):199–207.

Paulsen, C. M., and K. Wernstedt. 1995. Cost-effectiveness analysis for complex managed hydrosystems: An application to the Columbia River Basin. Journal of Environmental Economics and Management 28(3):388–400.

Perrings, C., K. G. Maler, C. Folke, C. S. Holling, and B. O. Jansson. 1995. Biodiversity Loss: Economic and Ecological Issues. Cambridge: Cambridge University Press.

Polasky, S., and A. R. Solow. 1995. On the value of a collection of species. Journal of Environmental Economics and Management 29(3):298–303.

Reilly, E. A. 1995. Multiple Criteria Decision Making: A Case Study of the Columbia River Basin Salmon Recovery Plan. Ph.D. dissertation. Pullman: Washington State University.

Rollins, K. 1997. Wilderness canoeing in Ontario: Using cumulative results to update dichotomous choice contingent valuation offer amounts. Canadian Journal of Agricultural Economics 45(1):1–16.

results to update dichotomous choice contingent valuation offer amounts. Canadian Journal of Agricultural Economics 45(1):1–16.

Shaw, W. D., and P. Jakus. 1996. Travel cost models of the demand for rock climbing. Agricultural and Resource Economics Review 25(2):133–142.

Shogren, J. F., and T. D. Crocker. 1995. Valuing ecosystems and biodiversity. Pp. 33–47 in Conservation of Great Plains Ecosystems: Current Science, Future Options. S. R. Johnson and Aziz Bouzaher, eds. Boston: Kluwer Academic.

Simon, B. M., C. S. Leff, and H. Doerksen. 1995. Allocating scarce resources for endangered species recovery. Journal of Policy Analysis and Management 14(3):415–432.

Smith, V. K. 1996. Estimating economic for nature: Methods for non-market valuation. New Horizons in Environmental Economics series. Cheltenham, United Kingdom: Elgar; distributed by Ashgate, Brookfield, Vermont.

Stevens, T. H., T. A. More, and R. J. Glass. 1994. Interpretation and temporal stability of cv bids for wildlife existence: A panel study. Land Economics 70(3):355–363.

Teclaff, L. A., and E. Teclaff. 1994. Restoring river and lake basin eco-systems. Natural Resources Journal 34(4):905–932.

Willis, K.G., G. D. Garrod, J. F. Benson, and M. Carter. 1996. Benefits and costs of the wildlife enhancement scheme: A case study of the pevensey levels. Journal of Environmental Planning and Management 39(3):387–401.

Willis, D. B., and N. K. Whittlesey. 1998. Water management policies for streamflow augmentation in an irrigated river basin. Journal of Agricultural and Resource Economics, 23(1):170–190.

Appendix G

Figures

Photograph of Glen Canyon Dam and Powerplant showing water release capacities of the powerplant, outlet works, and spillways.

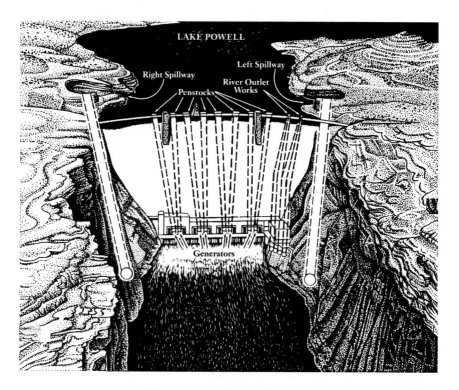

Glen Canyon Dam and Powerplant. From *The Colorado River Through Grand Canyon: Natural History and Human Change* by Steven W. Carothers and Bryan T. Brown. © 1991 The Arizona Board of Regents. Reprinted with permission by the University of Arizona Press.

DRAFT

Lake Powell Monitoring and Research Proposal[1]
Lake Powell Program Split Ad Hoc Group
November 17, 1998 (corrected copy)–to be on 12/8/98 TWG Agenda

Explanation of Change/Approach:

The following approach for identifying, developing, funding and conducting monitoring and research projects above Glen Canyon Dam (GCD) reflects:

1. the mandate within the Grand Canyon protection Act that the Adaptive Management Program (AMP) focus on effects of Secretarial actions on the Colorado River ecosystem downstream of GCD; and

2. the opportunity to provide adequate, long-term funding commitments for these AMP programs from outside sources.

All scientific activities related to Lake Powell would fit into one of the following categories with the specified funding sources:

WHITE AREAS: Those AMWG MO/INs that relate to downstream (below GCD) effects and conducted downstream of GCD:

- Funded by the AMP Budget
- Scope of work reviewed and approved by AMWG/TWG
- Includes all appropriate MO's and IN's
- GCMRC protocols apply (peer review, etc.)
- Accomplished by GCMRC and/or its contractors
- GCMRC will determine its capabilities to accomplish the work within funding personnel and other contraints

---

[1] Revised based on comments from Bruce Moore, Norm Henderson, Wayne Cook, Clayton palmer, Cliff Barrett, Bill Persons, Bill Davis, Rich Johnson, Barry Gold, Bill Vernieu, and Bob Winfree.

GRAY AREAS: Those AMWG MO/INs that relate to downstream effects, but conducted upstream of the dam:

- Part of the AMP and uses AMP procedures
- Funded by the Reclamation; e.g., O&M budget or other sources
- Scope of work developed by GCMRC and coordinated with USBR and the Lake Powell Group
- GCMRC protocols apply with PEP review before submission to MSWG/TWG
- Submitted to AMWG/TWG for review and recommended adoption
- Accomplished by CGMRC and/or its contractors

BLACK AREAS: Not directly related to downstream effects, conducted upstream of the dam:

- Funded by Reclamation, Lake Powell Group, or other sources
- Not part of AMP
- MO's and IN's are retained in a non-program information-desired category until next revision
- GCMRC protocols may not apply, cut data collection should be consistent for sharing of results
- Accomplished by USBR, participants in the Lake Powell Group, or others
- Results will be shared with GCMRC and AMWG

Note: GCMRC will present proposed budget split by 12/8/98

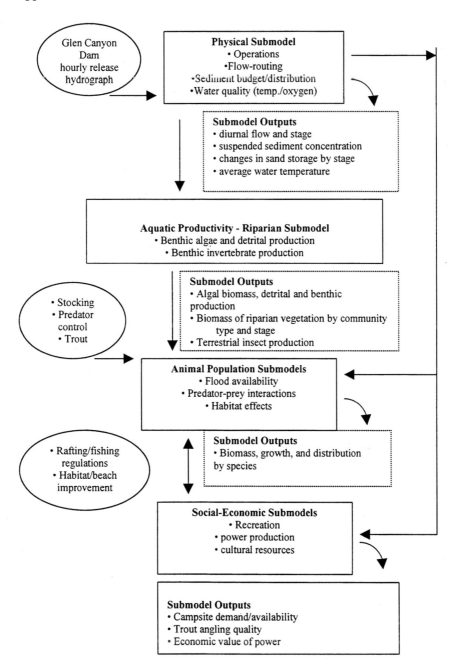

Figure 1 Overview of the structure of the proposed GCEM modeling shell.

Appendix H

Biographical Sketches
Committee on Grand Canyon Monitoring and Research

James L. Wescoat, Jr., is an associate professor of geography and member of the Institute of Behavioral Sciences at the University of Colorado, Boulder. Dr. Wescoat served on the National Research Council's Committee on the Future of Irrigation in the Face of Competing Demands. His research interests include the historical and cultural geography of water management in the western U.S., and the spatial logic of western water law, policies, and institutions. He has conducted comparative analyses of water policy issues in the Colorado, Indus, and Aral Sea basins. Dr. Wescoat received his M.A. and Ph.D. degrees in geography from the University of Chicago.

Trudy A. Cameron is a professor of economics at the University of California, Los Angeles. Her research interests include recreation economics and contingent valuation methods related to recreation and water resource management. She serves on the U.S. Environmental Protection Agency's Science Advisory Board's Environmental Economics Advisory Council. Dr. Cameron received her B.A. degree in economics from the University of British Columbia and her M.A. and Ph.D. degrees in economics from Princeton University. Dr. Cameron is also past vice-president of the Association of Environmental and Resource Economists.

Suzanne K. Fish is an associate professor of anthropology at the University of Arizona and the curator of the Arizona State Museum in Tucson, Arizona. Dr. Fish is especially recognized for her expertise in ethnobotany. Dr. Fish received her B.A. degree from Rice University, and her M.A. and Ph.D. degrees from the University of Arizona.

David Ford is the president of David Ford Consulting Engineers located in Sacramento, California. He is a lecturer at California State University, Sacramento, and at the University of California, Davis, and is a registered professional engineer in Texas, California, and Nevada. He has broad technical expertise and project experience in the areas of decision support systems, hydrologic engineering, water resource planning, natural resource policy analysis, hydropower operations and economics, and technology transfer. Dr. Ford received his B.S., M.S., and Ph.D. degrees in civil engineering from the University of Texas at Austin.

Steven P. Gloss is an associate professor of zoology at the University of Wyoming. Dr. Gloss is the former director of the Wyoming Water Resources Center and has served as president of the National Institutes for Water Resources and the Powell Consortium, a regional organization dealing with issues relevant to the Colorado River Basin. His research interests include water resources policy and management, aquatic ecology, fisheries science, limnology, and general ecology. He received his Ph.D. degree from the University of New Mexico in biology working on an interdisciplinary NSF-RANN project focusing on the Colorado Plateau and Lake Powell.

Timothy K. Kratz is a senior scientist at the Center for Limnology of the University of Wisconsin-Madison. His research interests include limnology, wetland ecology, and long-term dynamics of ecological systems. He served on the National Research Council's Committee to Assess EPA's Environmental Monitoring and Assessment Project, and he is currently serving on the Long-Term Ecological Research Network's Executive Committee. He received his Ph.D. degree in botany from the University of Wisconsin-Madison.

Wendell L. Minckley is a professor of biology at Arizona State University, with current research interests in conservation biology, aquatic ecology, and ecological and systematic ichthyology. He has published about 200 technical works and trained more than 50 graduate students in these areas of interest on aquatic systems and biota in the southwestern United States and northern Mexico. He received his B.S. degree in zoology from Kansas State University, his M.A. degree in zoology (ichthyology) from the University of Kansas, and his Ph.D. degree in biology (aquatic/radiation ecology; minor geology) from the University of Louisville.

Peter R. Wilcock is a professor in the department of geography and environmental engineering at The Johns Hopkins University. His research focuses on the mechanics of sediment transport and its application to problems of river erosion and sedimentation, on human impacts on river channel change, and on channel maintenance flows. In 1991–1993, Dr. Wilcock participated in an evaluation of trial reservoir releases for channel maintenance on the Trinity River, California. He received his B.S. degree in physical geography from the University of Illinois, Urbana-Champaign, and his M.S. degree in geomorphology from McGill University. Dr. Wilcock received his Ph.D. degree in geology from the Massachusetts Institute of Technology.

Jeffrey W. Jacobs is a senior program officer at the National Research Council's Water Science and Technology Board. His research interests include institutional and policy arrangements for water resources planning and international cooperation in water development. He has studied these issues extensively in Southeast Asia's Mekong River Basin and has conducted comparative research between the Mekong and the Mississippi River systems. Dr. Jacobs received his Ph.D. degree in geography from the University of Colorado.